Bacteriophages in Therapeutics

Edited by Sonia Bhonchal Bhardwaj

Published in London, United Kingdom

IntechOpen

Supporting open minds since 2005

Bacteriophages in Therapeutics
http://dx.doi.org/10.5772/intechopen.90988
Edited by Sonia Bhonchal Bhardwaj

Contributors
Derek Lin, Henry C. Lin, Ramasamy Palaniappan, Govindan Dayanithi, Igomu Elayoni Emmanuel, Ratnakar Deole, Chelsea Truitt, Amresh Kumar Singh, Vivek Gaur, Ankur Kumar, Cristina Paiva De Sousa, Andréa Cristina Bogas, Felipe De Paula Nogueira Cruz, Igor Vinícius Pimentel Rodrigues, Geusa Felipa De Barros Bezerra, Maria Do Desterro Soares Brandao Nascimento, Katia Regina Assuncao Borges, Sharon Shui Yee Leung, Wei Yan, Subhankar Mukhopadhyay, Kenneth K.W. To, Ayariga Joseph Atia, Abugri Daniel Azumah, Bedi Deepa, Derrick Dean, Sonia Bhonchal Bhardwaj, Seema Kumari

First published in London, United Kingdom, 2021 by IntechOpen
IntechOpen is the global imprint of INTECHOPEN LIMITED, registered in England and Wales, registration number: 11086078, 5 Princes Gate Court, London, SW7 2QJ, United Kingdom
Printed in Croatia

British Library Cataloguing-in-Publication Data
A catalogue record for this book is available from the British Library

Additional hard and PDF copies can be obtained from orders@intechopen.com

Bacteriophages in Therapeutics
Edited by Sonia Bhonchal Bhardwaj
p. cm.
Print ISBN 978-1-83962-212-0
Online ISBN 978-1-83962-213-7
eBook (PDF) ISBN 978-1-83962-228-1

Meet the editor

Dr. Sonia Bhonchal Bhardwaj, Ph.D., is a senior assistant professor at the Department of Microbiology, Dr. Harvansh Singh Judge Institute of Dental Sciences and Hospital, Panjab University, Chandigarh, India. She has published several international publications in reputed journals as well as books, book chapters, and congress proceedings. Dr. Bhardwaj has received grants from the Department of Science and Technology (India) and has worked on biofilm formation in *Enterococcus faecalis* in periodontitis, *Streptococcus mutans*, and phage therapy. She is a member of the Indian Association of Medical Microbiology (IAMM), Gastrointestinal Infection Society of India (GISI), Association of Microbiologists of India (AMI), and Society for Bacteriophage Therapy (SBRT).

Contents

Preface

Bacteriophages are widely used in food safety, agriculture, and different therapeutic areas to fight multidrug-resistant pathogenic bacteria. This book introduces bacteriophage morphology and biology, ecology, use of bacteriophages in oral, superficial, and systemic microbial diseases, tools in bacteriophage detection, and nanotechnology. Written by bacteriophage experts from all over the world, chapters highlight the theoretical and practical insights of this emerging field. This book is an essential guide and useful resource for clinicians, microbiologists, bacteriophage researchers, and dental and medical students.

I would like to thank the authors who contributed to this book. My wholehearted thanks go to Author Service Manager Sara Gojevic-Zrnic and Intech Open, pioneers in the field of medical and scientific publications, for taking up this work for publication.

Dr. Sonia Bhonchal Bhardwaj
Senior Assistant Professor,
Department of Microbiology,
Dr. Harvansh Singh Judge Institute of Dental Science and Hospital,
Panjab University, Chandigarh, India

Section 1

Introduction to Bacteriophages

Chapter 1

Bacteriophages: The Good Side of the Viruses

Igor Vinícius Pimentel Rodrigues,
Katia Regina Assunção Borges,
Maria do Desterro Soares Brandão Nascimento
and Geusa Felipa de Barros Bezerra

Abstract

Bacteriophages or phages are bacterial viruses that are known to invade bacterial cells and, in the case of the lytic phages, impair bacterial metabolism, causing them to lyse. Since the discovery of these microorganisms by Felix d'Herelle, a French-Canadian microbiologist who worked at Institut Pasteur in Paris, Bacteriophages begin to be used in the treatment of human diseases, like dysentery and staphylococcal skin disease. However, due to the controversial efficacy of phage preparations, and with the advent of antibiotics, commercial production of therapeutic phage preparations ceased in most of the Western world. Nevertheless, phages continued to be used as therapeutic agents (together with or instead of antibiotics) in Eastern Europe and in the former Soviet Union. Therefore, there is a sufficient body of data that incite the accomplishment of further studies in the field of phage therapy.

Keywords: Bacteriophages, therapy, antimicrobial, viruses, phages

1. Introduction

The resistance of pathogenic bacteria to most, if not all, currently available antimicrobial agents, has become a major problem in modern medicine, especially because of the increased numbers of immunosuppressed patients. The concern that humankind is approaching the "preantibiotics" era is becoming realer day by day, and this scenario increases the demand for the development of new antibiotics that can be used to treat these life-threatening diseases to human life [1].

Before the discovery and the wide spread use of antibiotics, it was suggested that bacterial infections could be prevented and/or treated with the administration of bacteriophages. Despite the fact that the clinical studies with bacteriophages were discontinued in United States and Western Europe, phages continued to be utilized in the former Soviet Union and in Eastern Europe. The results of the studies were extensively published in non-English journals, and, therefore, were not available to the western scientific community [1]. In this book chapter, we describe the history of bacteriophage discovery, the first clinical studies with phages, the application of phages in different bacterial diseases, the reason why its usage failed to prevail in the Western World, and last, but not less important, the future prospects of the use of Bacteriophages as therapeutical agents in bacterial diseases.

1.1 The discovery of bacteriophages and the first phage therapy research

Bacteriophages or phages are bacterial viruses that are known to invade bacterial cells and, in the case of lytic phages, impair bacterial metabolism, causing them to lyse. Since the discovery of bacteriophages, there has been a debate over claims for who really first discovered these microrganisms. Ernest Hankin, a British bacteriologist, reported in 1896 an antimicrobial activity against *Vibrio cholerae* in samples of water of Ganges and Jumna rivers in India, and he suggested that this phenomenon could be possible by the presence of an unidentified substance that passed through the fine porcelain filters and was heat labile, limiting the spread of cholera epidemics [2].

Two years later, the Russian bacteriologist Gamaleya observed a similar phenomenon while working with another bacterial species: *Bacillus subtilis* [3]. Other scientists also observed this event, but with other bacteria. However, none of them further explored their findings until Frederick Twort, a medically trained bacteriologist from England, reintroduced the subject almost 20 years after Hankin's observation by reporting a similar phenomenon and hypothesizing that it may have been due to, among other possibilities, a virus [4]. However, Twort did not continue his research because of many reasons, including financial difficulties [4–6] and only two years later, bacteriophages were "officially" discovered by Felix d'Herelle, a French-Canadian microbiologist at the Institut Pasteur in Paris [1].

Unlike Hankin and Twort, d'Herelle had almost no doubt about the nature of the observed phenomenon, and he proposed that it was caused by a virus capable of parasitizing bacteria. He and his wife Marie, on 18 October 1916, then decided to name this microorganism as "bacteriophage" [5]. The name derived from the words "bacteria" and "phagein" (to eat or devour, in Greek), implying that phages "eat" or "devour" bacteria. D'Herelle considered himself to be the discoverer of bacteriophages, but he acknowledged that his discovery was different from Twort's discovery. Also, in contrast to Twort, d'Herelle carried on studies of bacteriophages and strongly supported the idea that phages were live viruses – and not "enzymes" as many of his fellow researchers thought. The fight for the priority ceased eventually and many scientists accepted the independent discovery of bacteriophages, naming it as the "Twort-d'Herelle phenomenon" and later, the "bacteriophage phenomenon" [1].

2. First studies of phage therapy

After his discovery, d'Herelle used phage to treat dysentery, representing the first attempt to use bacteriophages to treat a bacterial disease. The study was conducted at the Hospital des Enfants-Malades in Paris in 1919 [5] under the supervision of Professor Victor-Henri Hutinel, the Hospital's Chief of Pediatrics. The phage preparation was ingested by d'Herelle, Hutinel and several hospital interns in order to test its safety before its usage by humans, more specifically, a 12-year-old-boy with severe dysentery. The patient's symptoms disappeared after a single administration of d'Herelle's antidysentery phage, and the boy fully recovered after a few days. The phage preparation proved its "efficacy" shortly after, when three other patients presenting bacterial dysentery that were treated with one dose of the preparation recovered within 24 hours of treatment [1].

However, the results of these studies were not published and the first reported application of phages used in the treatment of bacterial diseases happened only in 1921 in a study performed by Richard Bruynoghe and Joseph Maisin [7], who used bacteriophages to treat staphylococcal skin disease. The bacteriophages were injected into and around surgically opened lesions and it was observed a regression of the infections within 24 to 48 hours. In view of these promising results, several companies began commercial production of phages against various bacterial pathogens [1].

2.1 Marketing of phages

D'Herelle's commercial laboratory in Paris produced five phage preparations against various bacterial infections: Bacte-coli-phage, Bacte-rhinophage, Bacte-intesti-phage, Bacte-pyo-phage, Bacte-staphy-phage, and they were marketed by what later would become the large French company L'Oreal [5]. The production of therapeutic phages also began in the United States at that time. In the 1940s, the Eli Lilly Company (Indianopolis, Ind.) produced seven phages for human use against staphylococci, streptococci, *Escherichia coli*, and other bacterial pathogens, which consisted of phage-lysed, bacteriologically sterile broth cultures of the targeted bacteria (e.g., Colo-lysate, Ento-lysate, Neiso-lysate, and Staphylo-lysate) and the same preparations in a water-soluble jelly base (e.g., Colo-iel, Ento-iel, and Staphylo-jel). They were used to treat various infections, including abscesses, suppurating wounds, vaginitis, acute and chronic infections of the upper respiratory tract and mastoid infections. However, due to its controversial efficacy, and with the advent of antibiotics, commercial production of therapeutic phages ended in most of the Western World [8, 9]. Even so, phages continued to be used therapeutically (together with or instead of antibiotics) in Eastern Europe and in the former Soviet Union.

The institute, during its best times, employed approximately 1,200 researchers and support personnel, resulting in a production of phages of several tons a day, against a dozen bacterial pathogens, including *Staphylococci*, *Pseudomonas*, *Proteus*, and many enteric pathogens [1].

The bacteriophage laboratory of the Institute then began to produce phages for the treatment of many diseases, such as septicemia, furunculosis, and pulmonary and urinary tract infections and for the prophylaxis or treatment of postoperative and posttraumatic infections. In most of the cases, the phages were used against multi-drug resistant bacteria that were refractory to the conventional treatment with the majority of the antibiotics used in the clinical setting [10–16].

2.2 Experimental studies in animals

The first experimental studies that utilized animals in laboratories on the treatment of bacterial diseases using bacteriophages came from the Laboratory of William Smith and Smith and his colleagues [17–20] at the Institute for Animal Disease Research in Houghton, Cambrigeshire, Great Britain. In one of their first published papers, the authors reported the successful use of phages to treat *E. coli in vitro* infections in mice. In the next studies, [18–20] the authors found that a single dose of specific *E coli* phage reduced, by many orders of magnitude, the number of targeted bacteria in the digestive tract of calves, lambs, and piglets previously infected with a strain of *E. coli* that caused diarrhea. The treatment also ceased the associated fluid loss, and all the animals that were treated with the bacteriophages survived the bacterial infection. Furthermore, such positive results rekindled the interest in phage therapy in the West World and stimulated other researchers to investigate the possibility of using phages on the treatment of bacterial diseases caused by antibiotic resistant bacteria capable of causing human infections.

Another *in vivo* study performed by Soothill et al. [21] reported the importance of the phages in preventing and treating diseases induced experimentally in mice and guinea pigs infected with *Pseudomonas aeruginosa* and *Acinetobacter*, suggesting that its usage might be efficacious in preventing infections of skin grafts used to treat burn patients. However, it is uncertain if these "preclinical" studies preceded human clinical trials. Indeed, although many human trials were preceded by at least some *in vitro* studies using laboratory animals, the scientific literature regarding this topic is scarce.

Since the history of the discovery of the bacteriophages and some pioneer studies regarding this subject was already explored, the next section of this book chapter will explore the lytic and lysogenic cycles of phages, mode of action of these microrganisms when used in the therapy to treat bacterial diseases as well as some specific advantages and disadvantages in such use in the clinical settings.

2.3 Lytic and lysogenic life cycles of phages

Recent publications have provided interesting evidence that questions the notion that viruses are non-living organisms [22]. Erez et al., in their recent publication, identified a communication between viruses. They found a unique small-molecule communication system that controls lysis-lysogeny life cycles in a temperate phage [23]. Another study described the assembly of a nucleus-like structure during the viral replication of phage 201Φ2–1 in *Pseudomonas chlororaphis*, which suggested that phages have evolved a specialized structure to compartmentalize viral replication [24].

Phages can go through two different life cycles: the lytic and the lysogenic cycle. First, phages bind to the bacterial host specifically on a receptor found on the bacteria's surface and then injects its genetic material into the cell. The phage then takes advantage of the bacterium's biochemical machinery and replicate its genetic material, producing progeny phage. Subsequently, the phage synthesizes proteins such as endolysin and holin, which lyse the host cell from within. Holins are small proteins that accumulate in the cytoplasmic membrane of the host, allowing endolysin to degrade peptidoglycan and the progeny phage to escape the bacterial host. In the external environment, lytic phage can infect and destroy all bacteria nearby its initial bacterial host (**Figure 1**). The rapid proliferation and the large number of lytic phages are advantageous when they have therapeutic purposes. However, lytic phages have narrow host ranges and infect only specific bacterial species. Though, it can be overcame by giving a cocktail of different phages to patients afflicted by bacterial infections [25].

In the lysogenic cycle, the temperate phages do not immediately lyse the host cell, instead, they insert their genome into the bacterial chromosome at specific sites. This phage DNA now inserted into the host genome is called prophage, while the host cell containing the prophage is called a lysogen. The prophage then replicates along with the bacterial genome, establishing a stable relationship between them. The disadvantage of using temperate phage in phage therapy is that once the phage DNA is inserted into the bacterial genome, it can remain dormant or even alter the phenotype of the host [25].

Another advantage of using temperate phages in phage therapy is that the lysogenic cycle can continue indefinitely unless the bacteria are exposed to stress or adverse conditions. The signals that triggers such event vary from phage to phage, but prophage are commonly induced when bacterial stress responses are activated

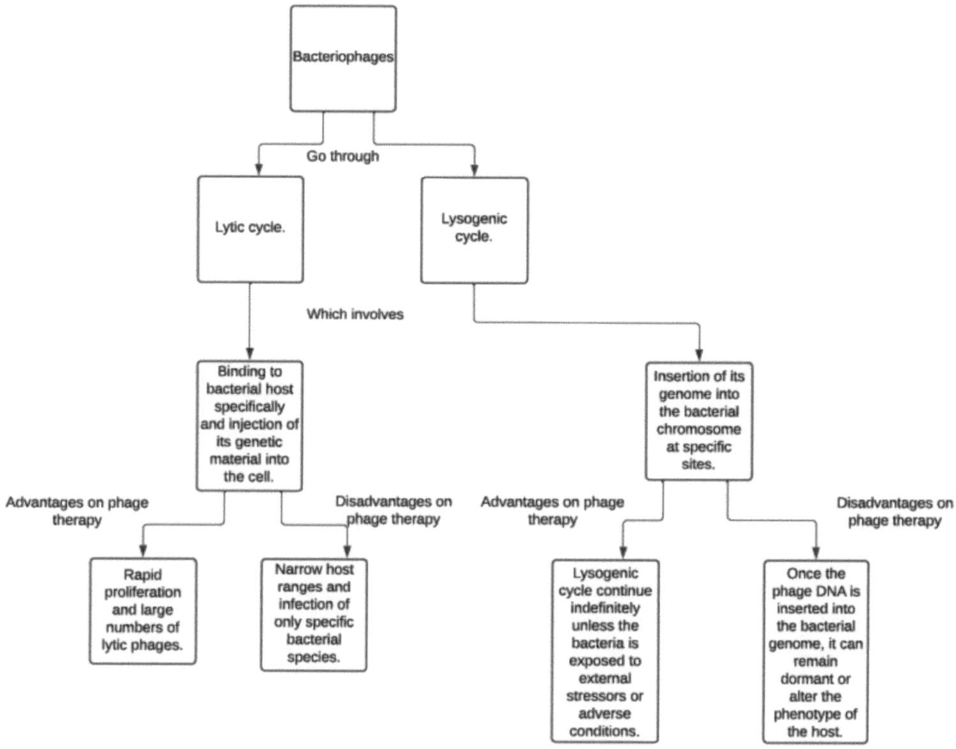

Figure 1.
Diagram representing the lytic and lysogenic cycle of the bacteriophages, as well as their advantages and disadvantages on phage therapy.

due to antibiotic treatment, oxidative stress, or DNA damage [26]. Once the lysogenic cycle finishes, expression of phage DNA starts and lytic cycle begins. In recent studies, it was found that phages that infect *Bacillus* species depends on small molecules called "arbitrium" to communicate to each other and make lysis-lysogeny decisions [23].

The biological implication of this phenomenon is very significant and explains why when phages encounters a large numbers of bacteria colonies, therefore, finding plenty of hosts to infect, they activate the lytic cycle. If host numbers is limited, the progeny phage then activates the lysogenic cycle and enters in a dormancy state. These recent findings stimulate other researches to be done to determine if there are other peptides also implicated in this phenomenon or if cross-talk is evident among different bacteriophage [25].

Furthermore, recent study regarding the full genetic sequence of the T4 phage (GenBank accession number AF158101) showed that the lysis of the bacteria by a lytic phage involves a complex process consisting several structural and regulatory genes. Besides, it is also possible that some therapeutic bacteriophages have some unique and unidentified genes or mechanisms responsible for effectively lysing their targeted bacteria. This led scientists to identify and clone, years later, an anti-*Salmonella* phage possessing a potent lethal activity against *Salmonella enterica* serovar *Typhimurium* host strains. Another study showed an unique mechanism for protecting phage DNA from the restriction-modification defenses of an *S. aureus* host strain. Further studies are necessary to gather information that are going to be useful to genetically engineer therapeutic phage preparations [27].

2.4 Mode of action of the bacteriophages

The first studies regarding the pharmacokinetics of bacteriophages showed that phages got into the bloodstream of laboratory animals after a single oral dose within 2 to 4 hours and that they were found in the following organs of the human body: liver, spleen, kidney, etc. in approximately 10 hours. Additionally, data concerning the period of time that the phages can remain in the human body indicate that it can happen for a long period of time, i. e., for up to several hours [28].

Despite the efforts in better understanding the pharmacokinetics of phages, their self-replication creates a complex scenario influenced by both decrease and proliferation. Although *in vivo* amplification of phages has been already performed, the topics are dominated by mathematical models of *in vitro* infections, which does not necessarily corresponds to *in vivo* amplification [29]. On the other side of it, phage lytic enzymes are considered as standard drugs in terms of pharmacokinetics. SAL200, a *S. aureus*-specific endolysin, has a $t_{1/2}$ between 0.04 and 0.38 hours after intravenous administration in healthy volunteers. The authors stated that, based on the molecular weight, renal clearance and drug distribution from the intravascular to the extravascular space should be minimal. Therefore, the presence of plasma proteases can explain the decay of this endolysin [30]. Other endolysins have a longer half-life (e.g., CF-302 has a half-life of 11.3 hours, while P128 has a half-life of 5.2 and 5.6 hours for the highet doses, 30 and 60 mg/kg, respectively) [31, 32]. Thus, as lytic enzymes in pre-clinical analyses shows an easier determination of its dosing regimen when compared to dosing regimen of phages, lytic enzymes are currently preferred to be used on patients [33]. In this sense, further studies are needed to better evaluate the pharmacological data concerning the lytic phages, including full-scale toxicological researches, before they can be used therapeutically in the West World [1].

2.5 Safety in the usage of phage preparations

From a clinical perspective, phages are apparently harmless. During the long period of usage of the phages as therapeutic agents in Eastern Europe and in the former Soviet Union (and before the antibiotic era, in the United States), phages have been administered to humans (i) orally, in tablet or liquid formulations (10^5 and 10^{11} PFU/dose) (ii) rectally (iii), locally (skin, eye, ear, nasal mucosa, etc.), in tampons, rinses and creams, (iv) aerosols or intrapleural injections, and (v) via intravenous access, though less frequently than the first four cited methods, and there are no reports of serious complications associated with their use [1].

Another aspect regarding safety of the bacteriophages usage is that they are extremely common in the environment (e. g., nonpolluted water has been reported to contain ca. $2x10^8$ bacteriophage per ml) [34] and are usually consumed in foods, highlighting their potential to be used as bioremediation agents on polluted environments. However, it would be prudent to ensure the safety of these microrganisms before using them as therapeutic agents, making sure, for example that: (i) they do not carry out generalized transduction and (ii) have genetic sequences possessing considerable homology with some genes related to antibiotic resistance, genes for phage-encoded toxins, and genes for other bacterial virulence factors [1].

2.6 Advantages in the use of bacteriophage therapy

Bacteriophage therapy presents many advantages such as high host specificity, preventing damage to normal intestinal flora, thus not infecting eukaryotic cells, low dosages required for the treatment, rapid proliferation inside the host bacteria, making them ideal candidates to treat bacterial infections [35]. Unlike antibiotics, another advantage in the usage of bacteriophages is that they reinfect the bacteria host and mutate alongside them [36].

However, high specifity of the phages can be both advantageous and a limiting factor. To use a monophage therapy it is necessary to check the efficacy of the phage by performing *in vitro* assays against the disease-causing bacteria before applying it in the patient, which can be a laborious task to do. The solution to this problem would be to use phage cocktails, which comprises a wide range of phages acting against different bacterial species or strains [37]. According to experts all around the world, an ideal phage cocktail consists of phages belonging to different families or groups so that it would target a broad range of hosts. Also, they would have to possess a high absorption ability to the highly conserved cell wall structures of the bacterial hosts. Additionally, the usage of phage cocktails may reduce the emergence of phage resistant bacterial population. On the other side, other researchers defend the sequential use of individual active phages to the patient, though, in clinical practice, it appears to be a difficult strategy to perform [38].

Not only bacteriophages *per se* can be used to treat bacterial infections. Their by-products can also do the trick. It was already reported that lytic enzymes showing function similar to lysozyme can also be used as an antibacterial agent or can be used in synergy with other antimicrobials like antibiotics to improve the efficacy of the treatment [39]. A phage derived protein, "endolysin", also possesses antibacterial and antibiofilm activity against ESKAPE pathogens [39–43]. V12CBD, a recombinant protein derived from bacteriophage lysine, PlyV12, was also able to attenuate virulence of *S. aureus* and also enhance its phagocytosis in mice [44].

2.7 Disadvantages in the usage of bacteriophage therapy

It is widely known that phages can be vector for horizontal gene transfer in bacteria, and in this process, bacteria can exchange virulence or antibiotic resistance gene, making these microrganisms resistant to a wide range of antibiotics [45]. Therefore, phages cannot harbor virulence factors or antibiotic resistance genes like integrases, site-specific recombinases, and repressor of the lytic cycle that may accelerate the integration of these genes in the bacterial hosts. Algorithms that can predict the mode of action of the phages as well as their virulent traits are available but their database needs to be constantly updated with a greater amount of genome sequence of phages [46].

Recent studies demonstrated an *in vivo* efficacy of phages against infections caused by ESKAPE pathogens (*Enterococcus faecium*, *Staphylococcus aureus*, *Klebsiella pneumoniae*, *Acinetobacter baumanni*, *Pseudomonas aeruginosa* and *Enterobacter* spp.), and their authors used fully characterized phages that showed no virulence factors or antibiotic resistance genes, therefore, they were considered safe as they do not provoked any allergic or immune response in the patient, and they were also stable at varied pH and temperature, making them ideal candidates for bacteriophage therapy [47–50].

Another limitation is the relatively weak stability of phages and their proper administration in order to reach the site of action. Phage preparations can be applied orally, nasally or topically [51, 52]. To overcome this limitation, studies were conducted and they have shown that phage's efficacy is improved when they are entrapped with liposomes [51, 53–55]. They can also reach the infection site in the form of a powdered formulation [56].

3. Future perspectives on phage therapy

There is an increasing urge to restock our ammunition of antimicrobials to combat the ever rising drug resistant bacterial pathogens. Effective antibiotic combinations are scarce and to add to the problem, the incoming of new drugs is also very low and happens in a very slow pace. Phages are a promising source of new antimicrobial drugs and they have been sparking up an interest on researchers all over the world, but still, their use is not approved on the United States and in Europe. But once limitations on their use is overcame, like preventing the phages to insert genes on their bacterial hosts that could confer them resistance to antibiotics and also the production of toxins, for example, the use of bacteriophages to treat bacterial diseases will be extremely helpful to treat patients affected by these bacterial diseases.

Acknowledgements

The authors of this book chapter would like to express their gratitude to the Federal University of Maranhao for providing the infra structure for us researchers to perform our experiments and also to FAPEMA (Foundation for the Support of Research and Scientific and Technological Development of Maranhao) for financing the experiments of the Laboratory of Mycology and the Cell Culture Laboratory (LCC) (Federal University of Maranhao – UFMA).

Conflict of interest

The authors declare no conflict of interest.

Author details

Igor Vinícius Pimentel Rodrigues[1*], Katia Regina Assunção Borges[2], Maria do Desterro Soares Brandão Nascimento[3] and Geusa Felipa de Barros Bezerra[3]

1 Post-Graduate Program of Adult Health, Federal University of Maranhão, São Luís, Maranhão, Brazil

2 Post-Graduate Program in Health Biotechnology by the Northeast Biotechnology Network, Federal University of Maranhao, Brazil

3 Department of Pathology, Federal University of Maranhão, São Luís, Maranhão, Brazil

*Address all correspondence to: igorviniciuspimentel@gmail.com

References

[1] Sulakvelidze A, Alavidze Z, J. Glenn MorriS J. BacterIophage Therapy. Antimicrob Agents Chemother. 2001;45(3):649-659.

[2] Hankin ME. The bactericidal action of the waters of the Jamuna and Ganges rivers on Cholera microbes . Ann. Inst. Pasteur 10:511-523 (1896). Bacteriophage. 2011;1(3):117-126.

[3] Samsygina GA, Boni EG. Bacteriophages and phage therapy in pediatric practice. Pediatriia. 1984;April(4):67-70.

[4] Twort FW. an Investigation on the Nature of Ultra-Microscopic Viruses. Lancet. 1915;186(4814):1241-1243.

[5] Summers WC. Félix d'Herelle and the Origins of Molecular Biology. 2nd Editio. New Haven, Conn.: Yale University Press; 1999. 248 p.

[6] Tolkacheva T V, Abakumov EM, Martynova VA, Golosova T V. Correction of intestinal dysbacteriosis with biological preparations in acute leukemia. Probl Gematol Pereliv Krovi. 1981;26(7):29-33.

[7] Bruynoghe R, Maisin J. Essais de thérapeutique au moyen du bacteriophage. CR Soc Biol. 1921;85:1120-1121.

[8] Eaton MD, Bayne-Jones S. Bacteriophage therapy: review of the principles and results of the use of bacteriophage in the treatment of infections. JAMA. 2015;23:1769-1939.

[9] Krueger AP, Scribner EJ. The bacteriophage: its nature and its therapeutic use. JAMA. 1941;19:2160-2277.

[10] Slopek S, Durlakowa I, Weber-Dabrowska B, Kucharewicz-Krukowska A, Dabrowski M, Bisikiewicz R. Results of bacteriophage treatment of suppurative bacterial infections. I. General evaluation of the results. Arch Immunol Ther Exp. 1983;31(3):267-291.

[11] Slopek S, Durlakowa I, Weber-Dabrowska B, Kucharewicz-Krukowska A, Dabrowski M, Bisikiewicz R. Results of bacteriophage treatment of suppurative bacterial infections. II. Detailed evaluation of the results. Arch Immunol Ther Exp. 1983;31(3):293-327.

[12] Slopek S, Durlakowa I, Weber-Dabrowska B, Dabrowski M, Kucharewicz-Krukowska A. Results of bacteriophage treatment of suppurative bacterial infections. III. Detailed evaluation of the results obtained in further 150 cases. Arch Immunol Ther Exp. 1984;32(3):317-335.

[13] SlopekS,Kucharewicz-KrukowskaA, Weber-Dabrowska B, Dabrowski M. Results of bacteriophage treatment of suppurative bacterial infections. IV. Evaluation of the results obtained in 370 cases. Arch Immunol Ther Exp. 1985;33(2):219-240.

[14] SlopekS,Kucharewicz-KrukowskaA, Weber-Dabrowska B, Dabrowski M. Results of bacteriophage treatment of suppurative bacterial infections. V. Evaluation of the results obtained in children. Arch Immunol Ther Exp. 1985;33(2):241-259.

[15] SlopekS,Kucharewicz-KrukowskaA, Weber-Dabrowska B, Dabrowski M. Results of bacteriophage treatment of suppurative bacterial infections. VI. Analysis of treatment of suppurative staphylococcal infections. Arch Immunol Ther Exp. 1985;33(2):261-273.

[16] S S, B W-D, M D, A K-K. Results of bacteriophage treatment of suppurative bacterial infections in the years

1981-1986. Arch Immunol Ther Exp (Warsz). 1986;35(5):569-583.

[17] Smith HW, Huggins MB. Successful treatment of experimental Escherichia coli infections in mice using phage: its general superiority over antibiotics. J Gen Microbiol. 1982;128(2):307-318.

[18] Williams Smith H, Huggins MB. Effectiveness of phages in treating experimental Escherichia coli diarhoea in calves, piglets and lambs. J Gen Microbiol. 1983;129(8):2659-2675.

[19] Williams Smith H, Huggins MB, Shaw KM. The control of experimental Escherichia coli diarrhoea in calves by means of bacteriophages. J Gen Microbiol. 1987;133(5):1111-1126.

[20] Williams Smith H, Huggins MB, Shaw KM. Factors influencing the survival and multiplication of bacteriophages in calves and in their environment. J Gen Microbiol. 1987;133(5):1127-1135.

[21] Soothill JS. Bacteriophage prevents destruction of skin grafts by Pseudomonas aeruginosa. Burns [Internet]. 1994;20(3):209-211. Available from: http://www.sciencedirect.com/science/article/pii/S1529104901000150%5Cnhttp://linkinghub.elsevier.com/retrieve/pii/S1438422105001293%5Cnhttp://www.ncbi.nlm.nih.gov/pubmed/17566713%5Cnhttp://linkinghub.elsevier.com/retrieve/pii/S0042682212004564%5Cnhttp://d

[22] Forterre P. To be or not to be alive: How recent discoveries challenge the traditional definitions of viruses and life. Stud Hist Philos Sci Part C Stud Hist Philos Biol Biomed Sci [Internet]. 2016;59:100-8. Available from: http://dx.doi.org/10.1016/j.shpsc.2016.02.013

[23] Erez Z, Steinberger-Levy I, Shamir M, Doron S, Stokar-Avihail A, Peleg Y, et al. Communication between viruses guides lysis-lysogeny decisions. Nature [Internet]. 2017;541(7638):488-493. Available from: http://dx.doi.org/10.1038/nature21049

[24] Chaikeeratisak V, Nguyen K, Khanna K, Brilot AF, Erb ML, Coker JKC, et al. Assembly of a nucleus-like structure during viral replication in bacteria. Science (80-). 2017;355(6321):194-7.

[25] Doss J, Culbertson K, Hahn D, Camacho J, Barekzi N. A review of phage therapy against bacterial pathogens of aquatic and terrestrial organisms. Viruses. 2017;9(3).

[26] Penadés JR, Chen J, Quiles-Puchalt N, Carpena N, Novick RP. Bacteriophage-mediated spread of bacterial virulence genes. Curr Opin Microbiol. 2015;23:171-178.

[27] Andriashvili IA, Kvachadze LI, Bashakidze RP, Adamiia RS, Chanishvili TG. Molecular mechanism of phage DNA protection from the restriction endonucleases of Staphylococcus aureus cells. Mol Gen Mikrobiol Virusol. 1986;8:43-45.

[28] Babalova EG, Katsitadze KT, Sakvarelidze LA, Imnaishvili NS, Sharashidze TG, Badashvili VA, et al. Preventive value of dried dysentery bacteriophage. Zh Mikrobiol Epidemiol Immunobiol. 1968;45(2):143-145.

[29] Brüssow H. Phage therapy: The Escherichia coli experience. Microbiology. 2005;151(7):2133-2140.

[30] Jun SY, Jang IJ, Yoon S, Jang K, Yu KS, Cho JY, et al. Pharmacokinetics and Tolerance of the Phage endolysin-based candidate drug SAL200 after a single intravenous administration among healthy volunteers. Antimicrob Agents Chemother. 2017;61(6).

[31] Cassino C, Murphy MG, Boyle J, Rotolo J, Wittekind M. Results of the

First In Human Study of Lysin CF-301 Evaluating the Safety, Tolerability and Pharmacokinetic Profile in Healthy Volunteers. 2014;209(9):2014. Available from: https://d1io3yog0oux5.cloudfront.net/_739ebeacddc4e1 0f31544124db5244b8/contrafect/db/257/1148/pdf/ContraFect+CF-301+ECCMID+2016+Poster.pdf

[32] Channabasappa S, Durgaiah M, Chikkamadaiah R, Kumar S, Joshi A, Sriram B. Efficacy of novel antistaphylococcal ectolysin P128 in a rat model of methicillin-resistant *Staphylococcus aureus* bacteremia. Antimicrob Agents Chemother. 2018;62(2).

[33] Abdelkader K, Gerstmans H, Saafan A, Dishisha T, Briers Y. The preclinical and clinical progress of bacteriophages and their lytic enzymes: The parts are easier than the whole. Viruses. 2019;11(2):1-16.

[34] Bergh Ø, Børsheim KY, Bratbak G, Heldal M. High abundance of viruses found in aquatic environments. Nature. 1989;340(6233):467-468.

[35] Domingo-Calap P, Delgado-Martínez J. Bacteriophages: Protagonists of a post-antibiotic era. Antibiotics. 2018;7(3):1-16.

[36] Pirnay JP, Verbeken G, Ceyssens PJ, Huys I, de Vos D, Ameloot C, et al. The magistral phage. Viruses. 2018;10(2):1-7.

[37] Chan BK, Abedon ST, Loc-Carrillo C. Phage cocktails and the future of phage therapy. Future Microbiol. 2013;8(6):769-783.

[38] Rohde C, Resch G, Pirnay JP, Blasdel BG, Debarbieux L, Gelman D, et al. Expert opinion on three phage therapy related topics: Bacterial phage resistance, phage training and prophages in bacterial production strains. Viruses. 2018;10(4).

[39] Lin DM, Koskella B, Lin HC. Phage therapy: An alternative to antibiotics in the age of multi-drug resistance. World J Gastrointest Pharmacol Ther. 2017;8(3):162.

[40] Viertel TM, Ritter K, Horz HP. Viruses versus bacteria-novel approaches to phage therapy as a tool against multidrug-resistant pathogens. J Antimicrob Chemother. 2014;69(9):2326-2336.

[41] Gong P, Cheng M, Li X, Jiang H, Yu C, Kahaer N, et al. Characterization of Enterococcus faecium bacteriophage IME-EFm5 and its endolysin LysEFm5. Virology [Internet]. 2016;492:11-20. Available from: http://dx.doi.org/10.1016/j.virol.2016.02.006

[42] Rios AC, Moutinho CG, Pinto FC, Del Fiol FS, Jozala A, Chaud M V., et al. Alternatives to overcoming bacterial resistances: State-of-the-art. Microbiol Res [Internet]. 2016;191:51-80. Available from: http://dx.doi.org/10.1016/j.micres.2016.04.008

[43] Zhang J, Xu L-L, Gan D, Zhang X. In Vitro Study of Bacteriophage AB3 Endolysin LysAB3 Activity Against Acinetobacter baumannii Biofilm and Biofilm-Bound A. baumannii. Clin Lab. 2018;64(6):1021-1030.

[44] Yang H, Xu J, Li W, Wang S, Li J, Yu J, et al. Staphylococcus aureus virulence attenuation and immune clearance mediated by a phage lysin-derived protein . EMBO J. 2018;37(17):1-15.

[45] Chen J, Novick RP. Phage-mediated intergeneric transfer of toxin genes. Science (80-). 2009;323(5910):139-41.

[46] McNair K, Bailey BA, Edwards RA. PHACTS, a computational approach to classifying the lifestyle of phages. Bioinformatics. 2012;28(5):614-618.

[47] Fish R, Kutter E, Wheat G, Blasdel B, Kutateladze M, Kuhl S.

Bacteriophage treatment of intransigent Diabetic toe ulcers: A case series. J Wound Care. 2016;25:S27–S33.

[48] Kishor C, Mishra RR, Saraf SK, Kumar M, Srivastav AK, Nath G. Phage therapy of staphylococcal chronic osteomyelitis in experimental animal model. Indian J Med Res. 2016;143(JANUARY):87-94.

[49] Wang Z, Zheng P, Ji W, Fu Q, Wang H, Yan Y, et al. SLPW: A virulent bacteriophage targeting methicillin-resistant *Staphylococcus aureus* in vitro and in vivo. Front Microbiol. 2016;7(JUN):1-10.

[50] Zhou W, Feng Y, Zong Z. Two new lytic bacteriophages of the Myoviridae family against carbapenem-resistant Acinetobacter baumannii. Front Microbiol. 2018;9(APR):1-11.

[51] Malik DJ, Sokolov IJ, Vinner GK, Mancuso F, Cinquerrui S, Vladisavljevic GT, et al. Formulation, stabilisation and encapsulation of bacteriophage for phage therapy. Adv Colloid Interface Sci [Internet]. 2017;249:100-133. Available from: http://dx.doi.org/10.1016/j.cis.2017.05.014

[52] Cooper CJ, Koonjan S, Nilsson AS. Enhancing whole phage therapy and their derived antimicrobial enzymes through complex formulation. Pharmaceuticals. 2018;11(2).

[53] Singla S, Harjai K, Katare OP, Chhibber S. Encapsulation of bacteriophage in liposome accentuates its entry in to macrophage and shields it from neutralizing antibodies. PLoS One. 2016;11(4):1-16.

[54] Chadha P, Katare OP, Chhibber S. Liposome loaded phage cocktail: Enhanced therapeutic potential in resolving Klebsiella pneumoniae mediated burn wound infections. Burns [Internet]. 2017;43(7):1532-1543. Available from: http://dx.doi.org/10.1016/j.burns.2017.03.029

[55] Chhibber S, Kaur J, Kaur S. Liposome entrapment of bacteriophages improves wound healing in a diabetic mouse MRSA infection. Front Microbiol. 2018;9(MAR):1-12.

[56] Chang RYK, Chen K, Wang J, Wallin M, Britton W, Morales S, et al. Proof-of-Principle Study in a Murine Lung Infection Model of. Antimicrob Agents Chemother. 2018;62(2):1-8.

Chapter 2

Oral Bacteriophages

Sonia Bhonchal Bhardwaj and Seema Kumari

Abstract

Bacteriophage or phage therapy involves using phages or their products as bio-agents for the treatment or prophylaxis of bacterial infections or diseases. Bacteriophages have the ability to regulate the oral microflora by lysing sensitive bacterial cells and releasing bacterial components with pro-inflammatory activity. Bacteriophages carry specific polysaccharide depolymerases that aid viral penetration and can disrupt the pathogenic process associated with biofilm and exopolysaccharide in the oral cavity. Oral diseases are mainly caused by biofilm forming microorganisms and phages are now being used for biocontrol of oral biofilms. Phages for *Actinomyces* species, *Aggregatibacter actinomycetemcomitans*, *Enterococcus faecalis*, *Fusobacterium nucleatum*, *Lactobacillus* species, *Neisseria* species, *Streptococcus* species, and *Veillonella* species have been isolated and characterized. Bacteriophages could be considered as potential therapeutic tools for the elimination of caries, periodontitis, and other diseases of the oral cavity.

Keywords: oral microbiome, oral phages, oral biofilms, oral diseases, bacteriophage therapy

1. Introduction

Bacteriophages are viruses that attack bacteria. Phages are now known to cure antibiotic-resistant bacterial infections as well as decrease bacterial virulence by overcoming the barriers bacteria used to avoid them. Bacteriophages are now being explored as potential therapeutic tools for the elimination of oral bacterial pathogens. Bacteriophages can disrupt pathogenic processes associated with biofilm and exopolysaccharide formation by oral microflora. Bacteriophages are a habitat to the human oral cavity where the oral pathogenic bacteria exist. Earlier studies show the isolation of oral bacteriophage from the oral cavity when an oral bacteriophage infecting *Lactobacillus casei* was obtained by Meyer et al. [1]. Subsequently, a range of oral bacteriophages infecting *Veillonella* species was isolated by Hiroki et al. in 1976, lytic bacteriophages for Actinomyces species were isolated by Tylenda et al. in 1985, oral bacteriophages specific for *Actinobacillus actinomycetocomitans* were described by Olsen et al. in 1993, oral bacteriophages specific for *Streptococcus mutans* were isolated by Delisle and Rotkwoski in 1993, and bacteriophages specific for *Enterococcus faecalis* were by Bachrach in 2003 [2–6]. Metagenomic analysis estimates 10^8–10^{10} virus-like particles existing per ml of human saliva and per gram of dental plaque [7]. The isolation studies for oral phages have been challenging, where phages have been obtained from clinical (saliva, plaque, oral washings) and environmental samples. The bacteriophages for oral bacteria implicated in various oral diseases have been described in the following section. The phages for the oral bacteria Actinomyces, Aggregatibacter, Fusobacterium, Parvimonas, Porphyromonas, Prevotella intermedia, *E. faecalis*, *S. mutans*, *Treponema denticola* are described here.

2. Actinomyces bacteriophages

Actinomyces species are found in healthy mouth but are also implicated in oral abcesses and oral-facial actinomycosis. Actinomyces, together with streptococci, initiates the biofilm development and formation of dental plaque [8]. Bacteriophages are used to block this co-aggregation to reduce the biofilm development without reducing health-related Actinomyces, which is part of the oral microbiome. The most commonly studied Actinomyces phage was AV-1, but it had a very narrow host range [9]. However, when the phage AV-1 was combined with AV-11, they lysed most of the indicator strains used for Actinomyces studies [10]. Actinomyces phages probably use surface structures of streptococci as receptors. These phages are from the families' Siphoviridae (61%) and Podaviridae (11%) [11, 12].

3. Aggregatibacter bacteriophages

Aggregatibacter is the causative agent of localized aggressive periodontitis. Aggregatibacter phages are mostly temperate phages and easy to isolate. Engineered Aggregatibacter bacteriophages that release biofilm degrading enzymes like dispersion B to breakdown biofilm have been used against periodontitis causing *Aggregatibacter actinomycetemcomitans* [13]. ϕAa 17 and Aaϕ23 are the most extensively studied Aggregatibacter phages [14]. These Aaϕ phages have a relatively broad host range. The limitation of these Aggregatibacter phages is that they can transfer antibiotic resistance genes, which are acquired macrolide lincosamide streptogramin B (MLS) resistance genes such as erm (A), erm (B), erm (C), erm (F), and erm (Q), and induce serotype conversion and release of leukotoxin [15]. In a recent study, it has been seen by metagenomics analysis that Aggregatibacter phages preferably lysogenize specific phylogenetic lineages not correlating with specific clinical conditions. They have either a very narrow host range or a broad host range [16]. The clinical conditions/impact in which these phages are used remains unknown.

4. Enterococcus bacteriophages

E. faecalis is one of the most frequently isolated species from nosocomial infections, endocarditis, bacteremia, urinary tract infections, meningitis, systemic infections. It has also been reported in periodontitis, which is a biofilm-mediated disease, tooth root infections, which are an example of endodontic biofilms, and also on implants. *E. faecalis* bacteriophages isolated belong to myoviridae and siphoviridae and are tailed phages. The bacteriophages isolated against *E. faecalis* strain of oral origin include phage IME-EF1 when administered intraperitoneally in a murine sepsis model protected the mice from lethal challenge around 60 to 80% mice surviving [17]. Another phage φEF 24C protected the BALB/C mouse model from the lethal challenge of *E. faecalis* [18]. Another phage EFDG1 tested on *E. faecalis* biofilms of post-treated root canal infections using an *ex vivo* two-chamber bacterial leakage model of human teeth showed dead bacteria in phage-treated teeth as compared to dentinal tubules of the control group [19]. The genetics of three phages ϕEF11, EFDG1, and EFLK1 has been studied by genome sequencing [20]. Full-genome sequencing of the EFDG1 genome revealed that it did not contain harmful genes and also efficiently prevented *E. faecalis* infection after root canal treatment. The authors concluded that phage therapy using these phages might be efficacious to prevent *E. faecalis* infection after root canal treatment. *E. faecalis* has also been recovered from periodontal pockets in 1–51.8% of chronic periodontitis patients [21]. In our recent study, a novel *E. faecalis* bacteriophage was isolated from sewage and was found effective in reducing biofilms formed by drug-resistant clinical isolates of *E. faecalis* from chronic periodontitis patients [22]. Passage of phage ϕ EF11 through *E. faecalis* strains JH2–2 harboring a defective prophage produced a new strain with more antimicrobial efficacy [23]. The use of enterococci bacteriophages can probably control colonization of teeth surfaces by reducing the biofilm in chronic periodontitis. The application of bacteriophages as a strategy to conventional antibiotic treatment particularly in the case of biofilm and multidrug-resistant strains is promising.

5. Streptococcus bacteriophages

The most important species that play a key role in dental plaque formation are oral Streptococci. The oral streptococci mainly constitute 12 species including *Streptococcus salivarius*, *S. agnisosus*, *S. mutans*, *S. constellates*, *S. cristareus*, *Streptococcus gordonii*, *S. mitis*, *Streptococcus oralis*, *S. parasanguis*, *Streptococcus pneumoniae*, *S. sanguis*, *S. sobrinus*. The initial colonizers of the tooth are *S. salivarius*, *S. sanguis*, *S. oralis*, and *S.gordonii*; however, *S. sobrinus* and *S. mutans* are more involved in dental infections [24]. Initial studies reported the isolation of lytic bacteriophages from human saliva [25]. The complete genome sequence of *S. mutans* lytic bacteriophage M102 was revealed [26]. In 2008, Van de Ploeg reported the complete genome sequence of prophage 15 infecting *S. gordonii*, which was a lysogenic phage [24]. A diverse group of around 50 bacteriophages that infect *S. mitis*, *S. mutans*, *S. oralis*, *S. salivarius*, and *S. sobrinus* have been identified and reported [27]. Unlike the *S. mutans* phages that are seen as lytic phages, the phages for *S. mitis* have been found as temperate. These temperate phages have the property to transfer host DNA into other bacterial strains. Seven phage-related gene clusters were detected in the genome of *S. mitis* B6, SM1, and ϕB6 prophages were isolated and sequenced [28]. Virulent pneumophages DP-1 and CP-1 were able to infect *S. mitis* and are also able to infect and replicate in commensal streptococci [29]. As *S. pneumoniae* and *S. mitis* carry numerous temperate phages in their genomes they are closely related and these virulent cross-infecting streptococcal phages and their enzymes are being used to biocontrol of oral infections [30].

6. Bacteriophages for oral anaerobes

6.1 Fusobacterium bacteriophages

Fusobacterium nucleatum bacteriophages have been isolated from saliva samples [31]. Siphovirus Fnpϕ02 could target three subspecies of *F. nucleatum*, *F. vincentii*, and *F. polymorphum*. The second phage Fnpϕ02 was rapidly absorbed on the cell surface but slow lysis was observed. In another study, non-infective phages were obtained by mitomycin C treatment of *F. nucleatum* [32]. The full-genome sequence and functional characterization of a novel lytic bacteriophage FNu1 against *F. nucleatum* which can break down oral biofilms have been reported recently [33].

6.2 Porphyromonas, prevotella, and tannerella

Prevotella phages have been detected *in vivo* [34]. Phages against *Porphyromonas gingivalis* and *Tannerella forsythia* have not been isolated so far. *P. gingivalis* that is an important anaerobic periodontal pathogen-causing microbial dysbiosis may protect itself in the periodontal pockets where many bacteriophages are preset by CRISPR-CAS systems providing it adaptive immunity [35]. These CRISPR-CAS systems are the only adaptive immune system in bacteria to fight phages/viruses, plasmids, transposons, integrative conjugative elements and are also found to target undesirable bacteria in the microbiome. On invasion or exposure to foreign DNA, the spacer sequences are transcribed into small CRISPR RNAs used by Cas proteins to cleave foreign DNA thus acquiring "acquired memory" of this adaptive immune system.

6.3 Treponema

A single study has reported the isolation of Treponema phage [36]. Phage ϕtd1 that belongs to Myoviridae family was harvested from the biofilm culture of *T. denticola* and its genome was detected by polymerase chain reaction.

6.4 *Veillonella* phages

It is a non-motile gram-negative diplococci. *Veillonella* is a part of the normal flora of the mouth is also associated with oral infections. Around 25 *Veillonella* phages have been isolated from mouth wash specimens. The small plaque-forming was found to be active against *Veillonella rodentium*. The large plaque formers were active against clinical *Veillonella* spp. isolates. Virion morphology was studied only for functional phages N2, N11, and N20 [37].

6.5 Lactobacillus

Bacteriophages for the caries associated with 12 strains of Lactobacillus including *L. casei* have been isolated. They have been divided into two groups: PL-1 is a lytic phage and temperate phage phi FSW of *L. casie* ATCC27139 [38].

7. Uses of oral bacteriophages

7.1 Bacteriophages and oral biofilms

The effectiveness of oral bacteriophages has been mainly seen by the reduction in the count of viable bacteria in the oral biofilms by using them. However, the phages were not able to reduce the amount of extracellular matrix in the biofilms [39]. Another factor while using phages is the phage therapy will be partially effective if particularly if the biofilm is old. The penetration and effect of phages on multispecies oral biofilms has also not been much studied. In a study in two species of biofilm constituting of phage-resistant and phage-susceptible bacteria, it was seen that the species composition of the biofilm may modulate phage effectiveness [40]. Limited studies show the application of oral phages *in vivo* using animal models. The efficacy of oral phages formulated in thermo-sustained release system against *E. faecalis* has been studied *in vivo* using a rat model. The study showed that per-apical inflammation of the tooth was improved after phage treatment [41].

7.2 Bacteriophages in oral diseases

Bacteriophages are being isolated to bacteria causing oral infections. Bacteriophages have been isolated to both aerobic and anaerobic microorganisms associated with periodontitis. Bacteriophages also constitute the majority of periodontal viral communities [42]. This variation in bacteriophages in healthy and periodontitis patients suggests a potential for more bacteriophage exploration. The use of bacteriophages has also been done in root canal treatment but targeted mainly against *E. faecalis*. Bacteriophages have also been explored for their therapeutic role in peri-implantitis [43] and also in the healing of oral mucosal infections [44].

7.3 Bacteriophages as antibiotic adjuvants

Phages can be used as adjuvants to antibiotic therapy. Resistance developed in phages can be reduced by using a cocktail of phages or phage recombinant lysins. Now, genetically engineered phages have also been developed to tackle resistance strains [45, 46].

The use of strictly lytic phages that infect only the target bacteria without affecting the normal microflora can be used as an alternative to local or systemic antibiotic therapy. This phage-based treatment can be designed in each case favoring personalized medicine.

8. Conclusion

The oral diseases caries, periodontal diseases, periapical and endodontic lesions, perimplantitis, and oral mucosal infections are microbial in origin. Bacteriophages are useful candidates for these biofilm-mediated diseases. As antibiotic resistance has become a matter of global concern, the bacteriophages or phage therapy can be used particularly to reduce the impact of acute infections. Moreover, antibiotics have a limited effect on the biofilm and are not much useful for the treatment of oral diseases. However, few bacteriophages are not effective against degrading biofilms; therefore, enzymatic or engineered phages are being investigated. Phages are low in cost, easy to isolate, and efficient against biofilm, and are bacteria specific. Phages have a great potential to be used in the prevention, control, and therapeutics of oral infections.

Conflict of interest

The authors declare no conflict of interest.

Author details

Sonia Bhonchal Bhardwaj[1]* and Seema Kumari[2]

1 Department of Microbiology, Dr. Harvansh Singh Judge Institute of Dental Sciences and Hospital, Panjab University, Chandigarh, India

2 Department of Microbiology, Panjab University, Chandigarh, India

*Address all correspondence to: sbbhardwaj2002@yahoo.com

References

[1] Meyer CE, Walters EL, Green LB. Isolation of a bacteriophage specific for a *Lactobacillus casei* from human oral material. Journal of Dental Research. 1958;**37**:175-178

[2] Hiroki H, Shiki A, Totsuka M, Nakamura O. Isolation of bacteriophages specific for the genus *Veillonella*. Archives of Oral Biology. 1975;**27**:261-268

[3] Tylenda C, Calvert C, Kolenbrander PE, Tylenda A. Isolation of Actinomyces bacteriophage from human dental plaque. Infection and Immunity. 1985;**49**:1-6

[4] Olsen I, Namork E, Myhrvold V. Electron microscopy of phages in serotypes of *Actinobacillus actinomycetemcomitans*. Oral Microbiology and Immunology. 1993;**8**:383-385

[5] Delisle AL, Rotkowski CA. Lytic bacteriophages of *S. mutans*. Current Microbiology. 1993;**27**:163-167

[6] Bachrach G, Leizerovici-Zigmond H, Zlotkin A, Naor R, Steinberg D. Bacteriophage isolation from human saliva. Letters in Applied Microbiology. 2003;**36**:50-53

[7] Naidu M, Robbes-Sikisaka R, Abeles SR, Boehm TK, Pride DT. Characterization of bacteriophage communities and CRISPR profiles from dental plaque. BMC Microbiology. 2014;**14**:175

[8] Mark Welch JL, Rosetti BJ, Rieken CW, Dewhirst FE, Borisy GG. Biogeography of a human oral microbiome at the micron scale. Proceedings of the National Academy of Sciences of the United States of America. 2016;**113**:E791-E800

[9] Delisle A. Growth of *Actinomyces viscosus* bacteriophage AV-1 in the presence of serum, saliva and dental plaque. Microbiology Letters. 1986;**33**:107-113

[10] Yeung MK, Kozelsky CS. Transfection of Actinomyces species by genomic DNA of bacteriophages from human dental paque. Plasmid. 1997;**37**:141-153

[11] Kolenbrander PE, Palmer RJ Jr, Periaswamy S, Jakubovics NS. Oral multispecies biofilm development and the key role of cell-cell distance. Nature Reviews. Microbiology. 2010;**8**:471-480

[12] Konoen E, Wade WG. Actinomyces and related organisms in human infections. Clinical Microbiology Reviews. 2015;**28**:419-442

[13] Szafranski SP, Winkel A, Stiesch M. The use of bacteriophages to biocontrol oral biofilms. Journal of Biotechnology. 2017;**250**(5):29-44

[14] Resch G, Kuik EM, Dietrich FS, Meyer J. Complete genomic nucleotide sequence of the temperate bacteriophage Aa phi23 of *Actinobacillus actinomycetemcomitans*. Journal of Bacteriology. 2004;**186**:5523-5528

[15] Stevens RH, de Moura Martins Lobo Dos Santos C, Zuanaff D, De Accioly Mattos MB, Ferreira DF, Kaxhlany SC, et al. Prophage induction in lysogenic *Aggregatibacter actinomycetemcomitans* cells co-cultured with human gingival fibroblasts and its effect on leucotoxin release. Microbial Pathogenesis. 2013;**54**:54-59

[16] Szafranski SP, Kiliam M, Yang I, Wieden GBD, Winkel A, Hegermann J, et al. Diversity pattern of bacteriophages infecting Aggregatibacter and Haemophilus species across clades and niches. The ISME Journal. 2019;**13**:2500-2522

[17] Khalifa L, Brosh Y, Gelman D, Coppenhagen-Glazer S, Beyth S, Poradusu-Cohen R, et al. Targeting *E. faecalis* biofilms with phage therapy. Applied Environment Microbiology. 2015;**81**:2696-2705

[18] Zhang W, Miz Yin X, Fan H, An X, Zhang Z, et al. Characterization of *E. faecalis* phage IME-EF1 and its endolysin. PLoS One. 2013;**8**:e80435

[19] Uchiyama J, Rashel M, Takemura I, Wakiguch H, Matsuzaki S. In silico and in vitro evaluation of bacteriophage phiEF 24 C, a candidate for treatment of *E. faecalis* infections. Applied Environment Microbiology. 2008;**74**:4149-4163

[20] Khalifa L, Coppenhagen-Glazer S, Shlezinger M, Kolt-Gutkowski M, Adini O, Beyth N, et al. Complete genome sequence of Enterococcus bacteriophage EFLK1. Genome Announce. 2015;**3**:e01308-e01315

[21] Sun J, Song X. Assessment of antimicrobial susceptibility of *E. faecalis* isolated from chronic periodontitis biofilm vs planktonic phase. Journal of Periodontology. 2011;**82**:84

[22] Bhardwaj SB, Mehta M, Sood S, Sharma J. Isolation of novel phage and targeting biofilms of drug resistant oral enterococci. Journal of Global Infectious Diseases. 2020;**12**:11-15

[23] Zhang H, Fouts DE, De Pew J, Stevens RH. Genetic modifications to temperate *E. faecalis* phage Ef11 that abolish the establishment of lysogeny and sensitivity to repressor and increase host range and productivity of lytic infection. Microbiology. 2013;**159**:1023-1035

[24] der Ploeg V Jr. Characterization of *Streptococcus gordonii* prophage PH15: Complete genome sequence and functional analysis of phage encoded integrase and endolysin. Microbiology. 2008;**154**:2970-2978

[25] Delisle AL, Rostkowski CA. Lytic bacteriophages of *Streptococcus mutans*. Current Microbiology. 1993;**27**: 163-167

[26] der Ploeg V Jr. Genome sequence of *Streptococcus mutans* bacteriophage M102. FEMS Microbiology Letters. 2007;**275**:130-138

[27] Delisle A. Bacteriophage-encoded enzymes for the treatment and prevention of dental caries and periodontal diseases. Patent US. 2004;**0234461**:A1

[28] der Ploeg V Jr. Genome sequence of the temperate bacteriophage PH10 from *Streptococcus oralis*. Virus Genes. 2010;**41**:450-458

[29] Queannane S, Leprohon P, Moineau S. Diverse-virulent pneumophages infect *Streptococcus mitis*. PLoS One. 2015;**10**:e0118807

[30] Garcia GP, Mendez FM, Garcia LE, Duz MR, De PA, Bastamante SN. Improved bactericidal enzybiotics against Pneumococcus and other bacteria. Patent. 2014:w02014191598A1

[31] Machua P, Daille L, Vines E, Berrocal L, Bittner M. Isolation of a novel bacteriophage specific for the periodontal pathogen *Fusobacterium nucleatum*. Applied Environment Microbiology. 2010;**76**:7243-7250

[32] Cochrane K, Manson McGuire A, Priest ME, Abouelleil A, Cerqueria GC, Lo R, et al. Complete genome sequences and analysis of *Fusobacterium nucleatum* subspecies animalis 7-1 bacteriophages Funu1 and Funu2. Anaerobe. 2016;**38**:125-129

[33] Mwila K, Teagan L, Brown JT. Genomic, morphological and functional characterization of novel bacteriophage FNu1 capable of disrupting *Fusobacterium nucleatum* biofilms. Scientific Reports. 2019;**9**:9107

[34] Pride DT, Salzman J, Relman DA. Comparisons of clustered regularly interspaced short palindromic repeats and viromes in human saliva reveal bacterial adaptations to salivary viruses. Environment Microbiology. 2012;**14**:2564-2576

[35] Chen T, Olsen I. *Porphyromonas gingivalis* and its CRISPR-Cas system. Journal of Oral Microbiology. 2019;**11**(1):1638196

[36] Mitchell HL, Dashper SG, Catmull DV, Paolini RA, Cleal SM, Slakeski N, et al. *Treponema denticola* biofilm-induced expression of a bacteriophage, toxin-antitoxin systems and transposases. Microbiology. 2010;**156**:774-788

[37] Totsuka M. Studies o *Veillonella* phages isolated from washings of human oral cavity. Bullentin Tokyo Medicine Dental University. 1976;**23**:261-273

[38] Meyers CE, Walter EL, Green LB. Isolation of a bacteriophage specific for a *Lactobacillus casei* from human oral material. Journal Dental Research. 1958;**37**:175-178

[39] Castillo-Ruiz IM, Vines ED, Monlt C, Fernandez J, Delgado JM, Hormamazabal JC, et al. Isolation of a novel *Aggregatibacter actinomycetemcomitans* serotype b bacteriophage capable of lysing bacteria within a biofilm. Applied Environment Microbiology. 2011;**787**:3157-3159

[40] Gonzalez CF, Domingo-Calap P. Phages for biofilm removal. Antibiotics. 2020;**9**:268. DOI: 10.3390/antibiotics 9050268

[41] Shlezinger M, Friedman M, Houri-haddad Y, Hazan R, Beyth N. Phages in a thermoreversible sustained release formulation targeting *E. faecalis* in vitro and in vivo. PLoS One. 2019;**14**(7):e029599. DOI: 10.1371/journal.pone.0219599

[42] Yu Z, Shan T-L, Li F, Yu T, Chen X, Deng X-T, et al. A novel phage from periodontal pockets associated with chronic periodontitis. Virus Genes. 2019;**55**:381-393. DOI: 10.1007/s11262-019-01658-Y

[43] Hashimoto K, Yoshinari M, Matsuzaka K, Shiba K, Inoue T. Identification of peptide motif that binds to surface of Zirconia. Dental Materials Journal. 2011;**30**:935-940

[44] Li GJ, Jiang DY, Zang X, Xu X. Keratinocyte growth factor phage model peptides can promote human oral mucosal epithelial cell proliferation. Oral Surgery Oral Medicine Oral Pathology Oral Radiology. 2013;**116**:e92-e97

[45] Santiago Rodriguez TM, Naidu M, Abeles SR, Bopehm TK, Ly M, Pride DT. Transcriptome analysis of bacteriophage communities in periodontal health and disease. BMC Genomics. 2015;**16**:549

[46] Tkhilashvili T, Wrinkler T, Muller M, Perka C, Trampuz A. Bacteriophages as adjuvant to antibiotics for the treatment of periprosthetic joint infection caused by multi-drug resistant *Pseudomonas aeruginosa*. Antimicrobial Agents Chemotherapy. 2020;**64**: e00924-e00919. DOI: 10.1128/AAC.00924-19

Chapter 3

Viruses of Extremely Halophilic Prokaryotes

Chelsea Truitt and Ratnakar Deole

Abstract

As viruses are known to be the most distinct source of biodiversity, it is not surprising that they are the most abundant biological group in hypersaline environments such as aquatic systems which have saturated salt concentrations. However, of more than 6000 known prokaryote viruses less than 100 are considered to be extremely halophilic (salt loving) and have the ability to infect bacteria. Combination of information obtained from culture dependent and culture independent methods allow better understanding of these viruses. This review will update the advances in halophilic viruses and its impact on the bacteriophage studies.

Keywords: halophiles, viruses, halophilic viruses, hypersaline environment, bacteriophages

1. Introduction

Halophiles are considered to be a part of a larger group of microorganisms called extremophiles. Like their name suggests, these microorganisms are able to thrive and survive within an extreme environment that would prove to be impossible for others. The extreme environment in which halophiles live are environments of high salinity or high salt concentration. Originally there were two categories of microorganisms considered to be halophiles, archaea and bacteria. However, within the last 50 years there was a discovery of yet another type of halophile, halophilic viruses. Viruses are in-fact one of the most abundant organism types within our biosphere and are able to infect organisms from all three domains of life [1–3]. Thousands of prokaryotic viruses have been identified but only a small portion of these are able to infect halophilic prokaryotes [4]. The origin of these halophilic viruses is not yet known but there are two differing hypotheses in regard to their arrival. One being that these viruses were the ones to originally give rise to other cell types; the other being that the different cell types gave rise to the viruses [5–11]. It has also been hypothesized that these halophilic viruses, also known as halophages, have served as a mode of genetic information between prokaryotes [9, 12]. This last hypothesis is supported by the fact that some of the largest viral sequences known, have some genetic similarities with the bacteria in which they prey upon. These halophages are able to be isolated from many different hypersaline environments all over the world and since their discovery have captured the interest of many different scientists. This is likely because not only do these viruses possess the ability to survive in these extreme environments, but they also have the ability to infect other halophilic organisms and live a wide variety of different life cycles. While there has been some

IntechOpen

progress on better understanding these viruses, further research is needed; to not only understand the effects they have had and continue to have on the environment in which they live but to also have a better idea of their potential uses across a wide variety of industries [9].

2. Discovery and initial studies of halophages

The discovery of the first halophilic virus happened accidentally while scientists were studying a known halophile, *Halobacterium salinarum,* which it had infected [9]. H. *salinarum* is an extremely halophilic archaea, despite its possibly misleading genus name of halobacterium. This archaeon is known for the discovery of bacteriorhodopsin, which is a light driven proton pump, that it utilizes as an energy source [13]. Since this discovery, there have been nine more viruses found that have the ability to infect halophilic bacteria as well as 56 other viruses that infect different species of halophilic archaea [9]. There has been further investigation into these types of viruses, including the work done by scientists Daniels and Wais. Working out of hypersaline ponds found in Jamaica they hypothesized that the samples collected both before and after rain fall would have a differing amount of both prokaryotes and halophilic viruses [14]. They thought that rain would act as a diluting agent in these ponds and would affect both the salinity of the water as well as its microbial community [9]. To evaluate this hypothesis, samples were collected both before and after rain fall. It was later learned that sample results not only varied between pre- and post-rain samples but also varied between large versus small sample sizes [9]. When Daniels and Wais were evaluating their "pre-rain", smaller sample volumes, they noted that there were fewer halophages as well as fewer plaques present [9]. The larger sample sizes under the same conditions on the other hand showed a larger number of viruses present as well as larger plaques [9, 15–17]. It has been suggested that plaque size is directly correlated with virulence, meaning, the larger a plaque appears to be the more virulent a virus is [14, 18, 19]. Not only the size of the plaques within the samples gives us information but also their opaqueness. Unlike the directly correlated relationship seen between virulence and plaque size, the opaqueness of the viral sample and its virulence are indirectly correlated. Meaning, the opaquer the plaques appear to be, the less virulent the viruses are, while samples that are clearer, suggests the viruses present are more lytic (or virulent) [20]. There have been arguments that these findings are not directly correlated with virulence and that the plaque appearance is due to chance. This is because of the argument that with the larger sample sizes there would be a larger virus to host ratio. This is supported by the consensus that more viruses tend to be present in comparison to prokaryotes within aquatic samples. Thus, an increase in the number of viruses' present could ultimately lead to more lysis taking place [9].

Another finding by Wais and Daniels was that after rainfall, there was a decrease in number of total halophilic prokaryotes present in the hypersaline ponds but an increase in free halophages [9]. It is important to note that extremely halophilic microorganisms generally thrive at a salinity of about 10%–30% NaCl. When concentrations drop below this range, this can disrupt the normal osmotic gradient present within the cell, resulting in cell lysis. Halophilic viruses on the other hand, are active at lower saline concentrations. From this information they suggested that the decrease in salinity lead to mortality in the prokaryotic population but prior to cell death, the viruses present within that host cell utilized the hosts cell machinery to replicate [9, 21–23]. While the viral population increased immediately after rainfall, this increase appeared to be short lived. Samples taken 24 days after the last rainfall showed that the viral population had decreased as the halophile population returned to baseline levels as well as the salinity of the water [9]. These findings led Wais and Daniels to hypothesize that the less-virulent strains of halophages can exist within these hypersaline environments, but they are not active when salt concentrations are too high. This means that when these prokaryotes are in their ideal hypersaline environments, they are safe from viral predation, but, once those water salt concentrations fall below ideal levels, they are more susceptible to active halophilic viral infections [9]. While these halophages may not be active at high salt concentrations, they are still present and living within the host cells genetic material as a prophage. When the salinity of the aquatic environment would decrease, due to instances such as rainfall, these same viruses would become active and utilize the hosts cell machinery prior to host cell death [9]. They believed this was the strategy employed by these viruses to ensure that they are able to remain stable despite the less favorable hypersaline environment. This type of relationship is viewed by some as mutually beneficial [24]. The prokaryotes present in the extreme hypersaline environment can live without the concern of viral predation, while the halophages are also able to co-exist within the host and make use of its cell machinery prior to inevitable cell death [9]. Similar studies and findings were conducted by Torvisk and Dundas while they were investigating their halophilic viral isolate, Hs-1 [25]. They observed that when infection of *Halobacterium salinarum* took place in a lower salinity atmosphere, the virus appeared to be more virulent. Conversely, when the viral infection took place in an environment of higher salinity the virus behaved in a more lysogenic fashion [9]. They also noted that the rate at which the virus could adhere to the surface of these prokaryotes also decreased with increasing salinity, suggesting that as salinity increases, adsorption decreases. The findings between the four previously mentioned scientists help further support their theories from both an environmental observation and laboratory point of view.

3. Halophilic viruses and their infection cycles

The process in which these viruses infect their host are similar to those of other viruses. Firstly, viruses adhere to the surface of the targeted cell and work their way into the cell's cytoplasm without detection. The next step in this process would be for the virus to replicate its genetic material using the hosts cell machinery. This process takes place in different fashions and timeframes that will be discussed later. The majority of halophilic viruses that have been discovered and studied contain DNA as their genetic material. There has yet to be a discovery of a halophilic virus that is made up of RNA [26]. After the virus replicates its genome and it is transcribed into mRNA, that mRNA is then translated into more viral particles. These particles then assemble and eventually leave the host cell to go on and infect another host. There can be variations of this process between different types of viruses and their own specific cycles. For example, viruses that behave in a lytic fashion proceed to take over the hosts cell machinery to produce its own progeny, resulting in destruction or lysis of the cell. Viruses that partake in this lytic lifestyle are also termed to be virulent. There are also viruses known as temperate viruses. This type of virus is known to infect halophilic prokaryotes in one of two ways. The first being that after invasion of the host cell they are able to integrate their own genetic information into the genome of the host and exist as a prophage. As the host cell undergoes its own replication cycle, the viral DNA is replicated along with it and is being passed along to daughter cells. Once the infected cell is subjected to instances of stress, this can cause the prophage to enter into the lytic cycle, take over the hosts cell machinery and eventually lead to cell lysis. The second type of temperate virus known to invade halophilic prokaryotes works by replicating its genetic material within the cytoplasm of the host cell similarly to a plasmid. Viruses are also able to partake in a chronic infection type lifestyle, also known as a persistent infection. This is where once the virus is inside the host cell it is able to continuously replicate its genetic material within the host without causing cell lysis. This type of lifestyle is best used to describe the non-lytic halophilic viruses. Examples of this type of virus would be the lemon-shaped virus His1 and also the pleomorphic virus, HRPV-1. How these viruses are able to leave the cell without causing cell lysis is not completely understood but it is hypothesized that it may be through budding of the plasma membrane [26].

4. Different Halophage morphologies

Not only can these halophages take part in multiple different types of infection cycles, but they are also known to have varying morphological structures. Of the halophilic viruses that have been isolated, there are 4 general categories of morphology in which they can be placed. The first being viruses with an icosahedral shape with either a medium length contractile tail, a short non-contractile tail or a long non-contractile tail. Second being viruses also with an icosahedral shape but this virus type also has an internal cell membrane. The final two categories of possible cellular morphologies include viruses that are pleomorphic and viruses that are lemon-shaped [26]. These varying morphologies can help us better understand the process in which these viruses are able to come into contact with the host cells and integrate their genetic information. In general, viruses are able to adhere to the hosts cell surface through some sort of receptor molecule. However, for haloviruses there has not been a specific receptor molecule identified. There has been a hypothesis that the receptor molecule for a known halovirus ϕCh1 could be a galactose residue found on the surface of a well-known haloalkaliphilic archaeon, *Natrialba magadii* [26]. The supporting evidence for this hypothesis is that the tail of ϕCh1 actually contains a protein fiber that has a galactose binding domain. Thus, when this virus comes within a close vicinity of the archaeon, its tail is able to bind with this galactose residue and allow the virus to invade the host cell. This theory is also supported by the fact that when there is a disruption in the genetic sequence of this virus that alters this protein fiber binding domain, the virus is no longer able to adhere to the hosts cell surface [26]. Furthermore, this would support the notion that if the virus has more than one protein fiber present within the tail, it could possibly have the ability to bind to more than one potential prokaryotic host [26].

Of the different viral morphology types, the tailed icosahedral type is one of the most common [27]. So naturally, there have been some tailed icosahedral halophilic viruses uncovered. These tailed viruses are thought to belong to either the *myoviridae* family or the *siphoviridae* family and preferentially infect archaea. These viruses not only have structural similarities between known bacteriophages, but they also have some genetic similarities. These findings suggest that they might share a common ancestor or that recombination might have taken place between the halophilic viruses and other cells found within the hypersaline environment. Examples of well-studied tailed icosahedral viruses include HSTV-2, HVTV-1 and HSTV-1. These three viruses are known to live a virulent life cycle and they do not contain an integrase sequence within their genetic material [26]. Integrase is the enzyme that lysogenic viruses generally use in order to integrate their own genetic information into the host cell's DNA to be replicated and transcribed. After these viruses infect the host cell and replicate their genetic information, they generally lyse the cell within 24 hours after the start of the infection. They also share an interesting feature that they are able to inactivate and reactivate their infectivity in response to lower salinities [26]. When the virus is subjected to a less than ideal or higher salt concentration, the active infection comes to a halt. When the virus is then placed back into a more optimal or lower salt concentration, the infection can resume. This type of adaptation is useful to viruses that are found in natural aquatic habitats because with rainfall or lack thereof, there can be either an increase or decrease in the water's salinity. This adaptation helps ensure that the virus can survive in instances of drought, where it may be a longer period of time before there is a decrease in the water's salt concentration. All halophilic viruses however, are not of the tailed icosahedral shape. There have also been halophilic viruses uncovered that have either the spherical or pleomorphic type morphology. These pleomorphic viruses can be isolated from not only aquatic habitats but salt crystals as well. They are also known to carry out a non-lytic lifestyle within their host. Examples of this type of virus would include His2 and HRPV-6 [26]. Finally, there are also the lemon-shaped viruses, which could possibly be the most common morphology for halophilic viruses.

5. Genetic studies of halophilic viruses

The genetic material that makes up the DNA of all organisms consist of four different nucleotides: Guanine, Adenine, Cytosine and Thymine. In the instance of RNA, the thymine is replaced with uracil. The natural pairing that takes place in double stranded DNA is that guanine pairs with cytosine through three hydrogen bonds and adenine binds with thymine through 2 hydrogen bonds. This hydrogen bonding plays an important role in not only the structure of DNA but also in the stability of the DNA and its resistance to denaturation. Genetic studies are a great way to not only better understand the organism in which you are studying but it is a great comparison tool to view similarities between both organisms of the same species and organisms from differing species. Studying genomics also allows scientists to form a road map of sorts that can help indicate which part of an organisms' genetic information is responsible for different actions and characteristics. Genetic studies of halophiles for example, help provide insight into how these organisms are able to survive in such harsh and hypersaline environments. Over the years, many different software programs have been designed to house known genetic information and to give scientists access to this information. One such program is called GAAS or "genome abundance and relative size". This program was originally developed to examine aquatic viruses on a global level. Viruses can be made up of both DNA and RNA that can be single stranded or double stranded. Viruses however, do not contain their own replication machinery and must utilize the machinery of their host to produce viral progeny. The GAAS program previously mentioned suggests that a large number of marine viruses have single stranded DNA as their genetic information type. However, most genomes of sequenced halophilic viruses appeared to be made up of double stranded DNA. According to GAAS the average size of the halophilic viral genomes were 51-263 kbps and the size of the viral genome appeared to be smaller when compared to that of its host. Another important aspect to take into consideration when studying genomics is whether or not portions of an organism's genome is "GC rich". Meaning that there is a higher percentage of guanine and cytosine nucleotide pairs in comparison to adenine and thymine base pairs. This information is important because due to the higher number of bonds between guanine and cytosine, it is thought that areas that are rich in these base pairs are considered to be more stable. Hence, there might be a higher GC concentration in certain areas within a organism's genome that encode for proteins that are vital for survival. If a protein is needed for survival, this would mean that specific portion of the genome would be conserved across most organisms within that species. When studying halophilic viruses and their GC content, it was observed that species with a specific GC content appeared to cluster together [9].

6. Examples of halophilic viruses

One of the first halophilic viruses to be analyzed at the molecular level was the virus ϕH. This temperate virus is known to infect archaea, specifically *Halobacterium salinarum*. ϕH's DNA is made up of double stranded DNA that is approximately 59kbp long and has a GC content percentage of 65% [9]. When it's genetic material was compared to that of it's host, there were very few similarities. Once infection with this virus takes place, the virus exists are a prophage in a closed circular state. This virus' structure was also observed to be very reliant on the salt concentration of its environment. Further investigation of this halophilic virus led to the discovery of 8 different variants, termed ϕH1- ϕH8 [9]. Another virus to be isolated from *Halobacterium salinarum* is the halovirus ϕN. ϕN has a icosahedral morphology with a non-contractile tail. Its DNA is made up linear double stranded DNA and is 56 kbps long with a GC content of 70% [28]. This virus is also known to carry out a virulent lifestyle.

The next two halophilic viruses were both isolated from a saltern in Australia and are named HF1 and HF2. These two viruses have quite a few characteristics in common. They both have the tailed capsid morphology and belong to the viral family *myoviridae*. Both of their genetic materials consist of double stranded DNA and they have the same 55.8% GC content present. Both viruses are also very sensitive to low NaCl concentrations and if there is no magnesium present, a minimum of 2 M NaCl is required for them to be active. Also, when these viruses infect their host they carry out a persistent infection type lifestyle. While these viruses appear to be very similar, there is a small difference in their genome size as well as differences in the halophilic prokaryotes they can infect. HF1 has a 76kbp sized genome and has three known halophilic hosts: *haloferax*, *halobacterium* and *haloarcula*. HF2 on the other hand, has a 77kbp long genome and is only known to be able to infect *Halobacterium sacchoravorum*. Further investigation into the genome of HF2 revealed that it does have some similarities between the genome of a well-known mesophilic bacteriophage. It's genome also exhibited some mosaicism which leads scientists to think that it's genetic material originated from a variety of different organisms that are not necessarily of halophilic viral origin.

The first and only haloalkaliphile type virus known to date is the tailed virus, ϕCh1. This virus is made up of linear double stranded DNA and is 58.5 kbps long with a GC content of 62%. This host works by invading host and integrating itself into the hosts genome. From there the virus is known to carry out a temperate infection lifestyle and requires the molarity of NaCl to be at least 2 M in order to remain active. While scientists were further investigating ϕCh1, this led to the discovery of the methyltransferase gene and its corresponding protein MϕCh1. It is also interesting to note that the protein products produced from this particular virus app appear to be acidic with isoelectric points ranging from 3.3–5.2 [9].

The next two halophilic viruses to be discussed were the first spindle shaped or lemon shaped halophilic viruses discovered. Naturally these two viruses do have a lot of similarities as well. They both are believed to be distantly related to the *fuselloviridae* family but they were given their own new family classification within the *salterproviridae* family [9]. They also share a common host, *Haloarcula hispanica*, which they both infect in a chronic and persistent fashion [9]. They are also noted to be not strictly lytic and are not lysogenic. Their genetic material is made up of double stranded DNA with a 39–40% GC content percentage [9]. An interesting feature of both viruses is also that they are able to encode for their own DNA polymerase which aides in their DNA replication within the host cell [27]. There is a slight difference in the sizes of these viruses, His-1 being 15 kbps long and His-2 being 16kbp long [9]. Another interesting discovery when further investigation the virus His-2, transfection experiments were carried out and this seemed to lead to broadening this viruses host range [9]. His-2 after the transfection experiments was then able to infect not only *Haloarcula hispanica* but also several *Haloferax*, *Halorubrum*, *Haloterrigena turkmenica*, and *Natrialba asiatica* species [9].

This next halophilic virus has been one of the most extensively studied viruses, SH1. This virus was isolated from a salt lake in Australia and it has a spherical morphology with a layered shell surrounded by a protein capsid that also has spikes [9, 24]. This structure is similar to other well-known mesophilic and hyperthermophilic phages such as PRD-1 – a gram negative phage, Bam35 – a gram positive phage, PBCV-1 – a algal virus and STIV – a hyperthermophilic virus [9]. Its genome is a total of 30kbp long and is made up of double stranded DNA [9]. This virus is known to have two different hosts, *Halorubrum sodomense* and *Haloarcula hispanica.*

HRPV-1 is the shortest halophilic virus known measuring at only 7 kbps long. Its genetic material is made up single stranded DNA with a GC content percentage of 54.2% [9]. It is also pleomorphic in morphology and is surrounded by a lipid envelope [9]. Its hosts consist of species within the *Halorubrum* genus. Another example of a short halophilic virus is the virus HHPV-1. It has a genome that is 8kbp in length with a GC content percentage of 56% [9]. Its structural morphology is also pleomorphic, and its DNA is a circular double stranded DNA [9]. The host for this virus is the archaeon *Haloarcula hispanica* [9].

Another identified halophilic virus is the spherical virus, SNJ1. This isolate was collected from its hosts within the *Natrinema* species [28]. It's genetic material is composed of circular double stranded DNA and is approximately 16.3 kbps long [25, 28]. Within its genetic sequence, 48.8–69.7% of that sequence is of GC content [28]. While this virus' genome has been further investigated it was noted that it is actually identical to a known plasmid pHH205 [29]. Another halophilic virus also isolated from species within the *Natrinema* genus is SNJ2. This virus is the first virus to be pleomorphic in regard to morphology that also carries out a temperate lifestyle [26]. The genetic material that makes up SNJ2 is a discontinuous double stranded circular DNA that is 16.9 kbps long with a GC content of 59.1% [30]. Further research into this virus suggests that the amount of SNJ2 viral particles present is dependent on the presence of the virus SNJ1 within the host. This is because SNJ1 is thought to act similarly to a plasmid, without which, ample amounts of SNJ2 is not produced [26].

7. The importance of culture-dependent and culture-independent studies of halophilic viruses

Within the field of microbiology, there are two different generalized approaches that can be used to study the microorganism(s) of interest. These being the culture-independent and culture-dependent approaches, both of which have their own advantages and limitations. The culture-independent approach is often used by environmental scientists that are interested in how microorganisms exist within their natural habitat. This type of information can tell us how these microbes respond to everyday natural elements that they are exposed to and help formulate questions as to how they are able to adapt to changes in environment. This approach can also give us insight into which microorganisms are able to coexist within the same environment and hypothesize what roles they may play with one another. While this approach can give scientists a large amount of useful information, there are some limitations. For example, if a mutation were to be observed across a species of microorganisms within a specific habitat, it would be difficult to pinpoint the specific cause. This is because in nature, there are countless amounts of influential factors to be taken into consideration.

The culture-dependent approach on the other hand, is an approach used by scientists within a laboratory setting. The microorganisms used to conduct these experiments are ones that have been harvested from a type of medium within the lab. Using this method, scientists are able to understand and better identify the microorganism of interest as well as observe its ability to adapt to changes initiated by researchers. Another advantage to this type of approach is that within the lab, microorganisms can be replicated and used on a larger scale. The culture-dependent approach is also important because this allows for isolation and purification of an organism of interest. Once a purified culture is obtained, meaning a culture is not contaminated by the presence of other organisms, it is able to be sent for genetic sequencing. As previously mentioned, the ability to sequence an organism's genetic material not only gives scientists more information as to why a certain organism may have certain behaviors or characteristics but it also opens an infinite number of doors for future possible research avenues. Just like the culture-independent approach, the culture-dependent approach also has limitations. For example, when how a microorganism may respond to a certain change within the environment this does not necessarily represent how it would respond in its natural habitat.

Examples of scientists utilizing the culture independent approach would be the scientists previously mentioned Daniels and Wais. They used this method to evaluate the fluctuation in the microbial communities within salt ponds due to rain fall. This type of experiment would not be able to be carried out within a laboratory setting with the same amount of accuracy. Another example of this approach being utilized was by scientist Oren et al. when they were collecting water samples from the Dead Sea over a specific time period. The purpose of this was to see how the microbial population as well as the amount of viral like particles fluctuated throughout the year. This also would be nearly impossible to carry out accurately within a laboratory setting. Many of the initial discoveries made of halophilic viruses were done with this culture-independent approach.

Examples of scientists utilizing the culture-dependent approach include studies done by Guixa-Boixareu et al. when they evaluated samples taken from a Spanish saltern [31]. Within the laboratory they were able to assess how varying salt concentration percentages affected the causes of prokaryotic death. It was found that at higher salt concentrations, more prokaryotes were terminated by amoebas or protozoans while at lower salt concentrations prokaryotic death was more due to viral lysis [9]. Scientists Torsvik and Dundas also used this approach to help support the environmental findings previously mentioned by Daniels and Wais. While they were studying the halophile *Halobacterium Salinarum* within the laboratory they were able to observe that the virulence exhibited by the virus Hs-1 appeared to be influenced by the saline concentrations present [9]. Their findings also suggested that as salt concentration is increased, the rate at which viruses are able to infect a host decrease.

In conclusion, both the culture-dependent and culture-independent methods of research are of vital importance within the scientific community. These approaches allow us to examine microbes within their natural habitats and from there formulate hypotheses about why they behave in the ways that they do. In the field of halophilic viruses, it is important that scientists continue to utilize both methods to gain a better overall understanding. There is still so much information needing to be collected and investigated to better understand these viruses and how they have influenced not only their environment but their possible influence on already known and studied bacteriophages.

8. Conclusion

Halophiles have long been an area of interest for many scientists. These organisms' abilities to withstand such harsh environments that would prove to be impossible to others, is one of their most intriguing characteristics. While halophilic bacteria and archaea have been on scientists' radar longer and have been more studied, this does not lessen the importance of the more recently discovered halophilic viruses. Both the culture-dependent and the culture-independent research strategies that have been used over the past 47 years have given us a great amount of insight into these halophilic viruses. We were able to learn not only where these types of viruses can be found but what types of prokaryotic hosts they prey upon. It has also helped us better understand the influential role that these viruses play within the environment. They help control and maintain the prokaryotic population, but they also have the potential to serve as a way of genetic communication between differing cell types. Haloviruses can appear in a wide variety of morphologies, each of which having their own life cycles and forms of infection. Each of these morphologies taken on by halophilic viruses can be found in hypersaline environments all over the world. It is undeniable that there has been many groundbreaking findings when looking into halophiles and the types of viruses than can infect them. While there is a decent amount of data present about halophilic viruses, more research as well as more in-depth research is needed. If we were able better understand the survival and infection methods employed by these halophilic viruses as well as their possible influential role on the genetics of other present bacteriophages, this would open up even more doors in the field of halophilic research.

Author details

Chelsea Truitt and Ratnakar Deole*
Department of Biochemistry and Microbiology, Oklahoma State University-Center for Health Sciences, Tulsa, Oklahoma, USA

*Address all correspondence to: ratnakar.deole@okstate.edu

References

[1] Oren, A (2002) Diversity of halophilic microorganisms: environments, phylogeny, physiology and applications. *J Ind Microbiol Biotechnol* 28: 56-63

[2] Oren, A (2009) Saltern evaporation ponds as model systems for the study of primary production processes under hypersaline conditions. *Aquat Microb Ecol* 56: 193-204

[3] Wommack KE, Colwell RR (2000) Virioplankton: viruses in aquatic ecosystems. *Microbiol Mol Biol Rev* 64(1):69-114

[4] Oren, A (2014) Taxonomy of halophilic Archaea: current status and future challenges. *Extremophiles* 18: 825-834

[5] Andersson SG, Kurland CG (1998) Reductive evolution of resident genomes. *Trends Microbiol* 6(7):263-268

[6] Forterre P (2006) The origin of viruses and their possible roles in major evolutionary transitions. *Virus Research* 117(1):5-16

[7] Forterre P, Prangishvili D (2009) The origin of viruses. Res Microbiol 160(7): 466-472

[8] Koonin E, Senkevich T, Dolja V (2006) The ancient virus world and evolution of cells. *Biology Direct* 1:29

[9] Sabet, S. (n.d.). Halophilic Viruses. *Advances in Understanding the Biology of Halophilc Microorganisms,* 81-116. doi:10.1007/978-94-007-5539-0_4

[10] Weinbauer MG, Rassoulzadegan F (2004) Are viruses driving microbial diversification and diversity? *Environ Microbiol* 6(1):1-11

[11] Wixon J (2001) Featured organism: reductive evolution in bacteria: *Buchner asp., Rickettsia prowazekii and Mycobacterium leprae*. Comp Funct Genomics 2(1):44-48

[12] Filee J, Chandler M (2010) Gene exchange and the origin of giant viruses. *Intervirology* 53(5):354-361

[13] Eichler, J. (2019). Halobacterium salinarum. *Microbe of the Month, 27*(7), 651-652.

[14] Lipton HL (1980) Persistent Theiler's murine encephalomyelitis virus infection in mice depends on plaque size. *J Gen Virol* 46(1):169-177

[15] Daniels LL, Wais AC (1984) Restriction and modification of halophage S45 in halobacterium. *Curr Micribiol* 10(3):133-136

[16] Daniels LL, Wais AC (1998) Virulence in phage populations infecting Halobacterium cutirubrum, FEMS *Microbiol Ecol* 25(2):129-134

[17] Wen K, Ortmann AC, Suttle CA (2004) Accurate estimation of viral abundance by epifluorescence microscopy. *Appl Environ Microbiol* 70(7):3862-3867

[18] Ramsingh AI, Caggana M, Ronstro, S (1995) Genetic mapping of the determinants of plaque morphology of coxsackievirus B4. *Arch Virol* 140(12):2215-2226

[19] Schloer GM, Hanson RP (1968) Relationship of plaque size and virulence for chickens of 14 representative Newcastle disease virus strains. *J Virol* 2(1):40-47

[20] Maloy SR, John E, Cronan J, Freifelder D (1994) Microbial genetics. Jones and Bartlett, Massachusetts, p 95

[21] Daniels LL, Wais AC (1990) Ecophysiology of bacteriophage

S5100 infecting halobacterium cutirubrum *Appl Environ Microbiol* 56(11):3605-3608

[22] Wais AC, Daniels LL (1985) Populations of bacteriophage infection Halobacterium in a transient brine pool. *FEMS Microbiol Ecol* 31:323-326

[23] Torsvik T. Dundas I (1980) Persisting phage infection in Halobacterium salinarium str. 1. *J Gen Virol* 47(1):29-36

[24] Porter, K., Kukkaro, P., Bamford, J.K., Bath, C., Kivela, H.M., Dyall-Smith, M.L.. and Bamford, D.H. (2005) SH1: a novel, spherical halovirus isolated from ab Australian hypersaline lake. *Virology* 335: 22-33

[25] Porter, K., Tang, S.L., Chen, C.P., Chiang, P.W., Hong, M.J., and Dyall-Smith, M. (2013) PH1: an archaeovirus of Haloarculaa hispanica related to SH1 and HHIV-2. *Archaea* 456318

[26] Atanasova, N. S., Bamford, D. H., & Oksanen, H. M. (2016). Viral-host interplay in high salt environments. *Environmental Microbiology Reports,* 431-444. doi:10.1111/1758-2229.12385

[27] Oren, A., Bratbak, G., and Heldal, M. (1997) Occurrence of virus-like particles in the Dead Sea. *Extremophiles* 1:143-149

[28] Luk AW, Williams TJ, Erdmann S, Papke RT, Cavicchioli R. Viruses of haloarchaea. *Life (Basel)*. 2014;4(4):681-715. Published 2014 Nov 13. doi:10.3390/life4040681

[29] Ziqian Zhang, Ying Liu, Shuai Wang, Di Yang, Yichen Cheng, Jiani Hu, Jin Chen, Yunjun Mei, Ping Shen, Dennis H. Bamford, Xiangdong Chen, 2012. Temperate membrane-containing halophilic archaeal virus SNJ1 has a circular dsDNA genome identical to that of plasmid pHH205. Virology, Volume 434(2):233-241 https://doi.org/10.1016/j.virol.2012.05.036.

[30] Liu, Y., Wang, J., Liu, Y., Wang, Y., Zhang, Z., Oksanen, H.M., Bamford, D.H. and Chen, X. (2015), Haloarchaeal temperate pleolipovirus SNJ2. Molecular Microbiology, 98: 1002-1020. https://doi.org/10.1111/mmi.13204

[31] Diez B, Anton J, Guix-Boixereu N, Pedros-Alio C, Rodriguez-Valera F (2000) Pulse-field gel electrophoresis analysis of virus assemblages present in a hypersaline environment. Int Microbiol 3(3):159-164

Section 2

Bacteriophage Therapy

Chapter 4

Light and Phages on Tackle of Infectious Diseases

Felipe de Paula Nogueira Cruz, Andréa Cristina Bogas and Cristina Paiva de Sousa

Abstract

There has been an important increase in the emergence of resistance in microbial population worldwide. This trajectory needs, necessarily new approaches to treat infectious diseases. The ability to detect and prevent the evolutionary trajectories of microbial resistance would be of value. Photodynamic inactivation (PDI) represents an efficient alternative treatment for diseases caused by viruses, which can cause infections well documented in various mammals. PDI can kill cells after exposure with the appropriate photosensitizer (PS), light of adequate wavelength combined with the presence of oxygen, without inducing resistance. Cytotoxic reactive species formed interaction with vital biomolecules leading to irreversible microbial inactivation. Bacteriophages can act on delivering antimicrobial agents into bacteria, which consist in a likely instrument for the treatment of infectious diseases. Non-enveloped bacteriophages are more difficult to tolerate photoinactivation than enveloped phages, which makes them an important model tool to evaluate the efficiency of PDI therapy against viruses that cause diseases in humans. Combination of photosensitizers and bacteriophage therapy can be employed to eradicate biofilms, contributing to control of infections also caused by drug-resistant bacteria.

Keywords: bacteriophages, biofilm, microbial resistance, photodynamic therapy, reactive oxygen species

1. Introduction

Despite the remarkable progress in human medicine, infectious diseases of microbial origin are one of main global concern to public health [1] worldwide. The relative unavailability of efficient drugs the misuse and/or excessive use of antimicrobials, are some factors that make infections harder or impossible to treat, increasing the risk of spreading diseases and deaths [2]. The gap in the discovery of new antibiotics over the decades [3, 4] also contributes to the increased risk of infectious diseases.

The emergence of antibiotic-resistant "superbugs" and their rapid global spread are alarming [5]. These microorganisms are members of a group known as nosocomial ESKAPE (*Enterococcus faecium*, *Staphylococcus aureus*, *Klebsiella pneumoniae*, *Acinetobacter baumannii*, *Pseudomonas aeruginosa* and *Enterobacter* spp.), associated with major risk of mortality in immunocompromised patients [6, 7]. Other highlights in the global list of drug-resistant priority pathogens are the third generation cephalosporins (3GC) resistant *Escherichia coli*, fluoroquinolone-resistant *Neisseria gonorrhoeae*, *Streptococcus pneumoniae*, *Salmonella* spp. and *Candida auris* [8, 9].

Drug-resistant pathogens can be transmitted through the hospital environment, increasing the severity in relation to Health Care-Associated Infections (HAIs) [10], also causing an important economic impact especially in developing countries. In these countries, the infectious diseases are more prevalent, and the prevention measures requires the use of drugs that maximize costs [7]. Given this scenario, effective antibiotics and strategies to combat antimicrobial resistance, prevent the high number of deaths each year and an economic crisis worldwide become urgent [11, 12]. This chapter describes photodynamic treatment, bacteriophages utilization and the combination of both as alternative therapies for minimize the excessive exposure of patients to antibiotic and risks of multi-resistant strains development.

2. Photodynamic therapy: past

The therapeutic potential of light has been used for hundreds of years by the ancient civilizations in Egypt, China, and India. Also, over 3000 years ago, light was used in conjunction with reactive chemicals to treat various conditions such as vitiligo, psoriasis, and some types of skin cancer. In China, it was introduced by Lingyan Tzu-Ming in the first century b.C., and, four centuries later, became a ritual practice in which it was based on exposing a piece of green paper containing a red dye and exposed to sunlight, then it was soaked in water and ingested right after [13, 14].

In the last few decades, Photodynamic therapy (PDT) has emerged as a promising intervention treatment for cancer therapy. However, it is widely used in the removal of small vessels and in the treatment of microbial infections [15]. Still, the first concepts of the nature of light emerged in the 17th century. Preliminary work on the properties of light, such as that of Christiaan Huygens, who used wave theory to explain the reflection and refraction of light in 1690, and, later, the discovery of the properties of electricity and magnetism, in the early 19th century [16, 17].

In fact, quantum theory started when Max Planck, in 1900 published an article that explained the spectral distribution of Blackbody Radiation, which perfectly fitted the laws of thermodynamics with the laws of electromagnetism. And in the same year, Oscar Raab was scientifically proven to have the beneficial effects of light. In his experiment, it was observed that the combination of light with the acridine dye (**Figure 1**) was lethal for Paramecium species. In the same year, the French neurologist, Jean Prime, discovered that oral eosin, used to treat patients with epilepsy, could cause dermatitis when exposed to sunlight [16, 18].

According to electrodynamic theory, light consists of an oscillating electromagnetic field that propagates as a wave through a vacuum or through a medium [16]. This means that when light propagates through space, it behaves like a wave, while when interacting with matter, it behaves like particles [16, 17, 19]. This concept was described by Einstein in 1905 based on the theories of Planck and Hertz for the explanation of the photoelectric effect (Eq. (1)). For that, Einstein assumed that light had

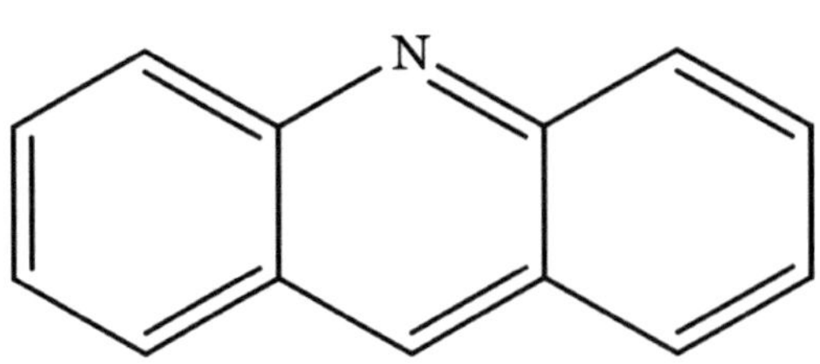

Figure 1.
Chemical structure of acridine.

a corpuscular nature, that is, it would be formed by small bundles of energy (quanta) called photons. Einstein also proposed the existence of a dependency relationship between the photoelectric emission and the frequency of incident radiation. For this theory to be valid, light could not be considered as a wave, but as a particle [17, 20].

Finally, in 1924, de Broglie created the hypothesis of wave-particle duality, which was soon recognized by Erwin Schrödinger who developed the wave propagation equation in matter in 1926 (Eq. (2)). Still during this period, other important scientists contributed to the establishment of quantum mechanics such as, for example, Max Born (Matrix Quantum Mechanics), Paul Dirac (Movement of sub-atomic particles), Werner Heisenberg (Uncertainty Principle) and Wolfgang Pauli (Principle of Exclusion) [20].

Equation 1: According to Einstein, each photon has an energy proportional to the frequency of light.

$$E_{photon} = hc / \lambda \quad (1)$$

E = de um quantum energy of light
h = Planck constant = 6,63 x 10–34 J.s
c = Speed of light (3×10^{10} cm/s)
λ = Frequency of light (Hz)

Equation 2: Erwin Schrödinger allows to determine to find the wave function of a particle, from the knowledge of the potential energy to which it is submitted.

$$-\frac{\hbar^2}{2m}\nabla^2\Psi + V\Psi = i\hbar\frac{\partial\Psi}{\partial t} \quad (2)$$

ħ = Planck constant (6,63 x 10–34 J.s) squared reduced
m = Particle mass
∇ = Ψ Laplacian – Spatial variation of the wave function
Ψ = Wave function
V = V potential that acts on the particle
i = Imaginary number given by the square root of −1
ət = Variation of wave function Ψ over time

In the same year, Policard [21] conducted a study where he detected the presence of porphyrins in high concentrations in malignant tumors. These, completely non-toxic, were able to destroy the tumor tissue in the presence of visible light and oxygen [21].

Later, in 1950 Schwartz demonstrated that the long-lasting phototoxic effect was not promoted only by hematoporphyrin. The action occurred due to an oligomeric mixture together with it. Since hematoporphyrin is eliminated quickly from the body, Schwartz enriched the oligomer mixture and this preparation was called hematoporphyrin derivative (HpD), which contains in addition to monomers, oligomers containing two to nine units of porphyrin [14, 22].

In the study conducted by Weishaupt [23], it was demonstrated that the destruction of tumor cells was due to the formation of singlet oxygen molecules.

Finally, in 1993, Photofrin ® (Axcan Pharma Inc., Canada) was approved for the treatment of superficial bladder cancer by the Canadian Health Protection Branch. Subsequently, in 1998 the Food and Drug Administration (FDA) authorized PDT in the treatment of cancer [24].

3. PDT fundamentals

PDT consists of a photochemical reaction between a photosensitizing agent and the oxygen that selectively destroys the target tissue, constituting an alternative modality clinically approved by several health agencies in many countries [25–29]. The photodynamic effect consists of causing a powerful and sustained photochemical reaction between light at a given wavelength, the photosensitizer (PS) and oxygen in the target tissue. Consequently, after the irradiation of light, PS converts O_2 into cytotoxic reactive oxygen species (ROS), where cell death can occur through mechanisms such as apoptosis, necrosis, or autophagy (**Figure 2**). However, recent studies have demonstrated the existence of other mechanisms with characteristics of necrosis and apoptosis. These new pathways of cell death, collectively called regulated necrosis, include a variety of processes triggered by different stimuli [14, 30].

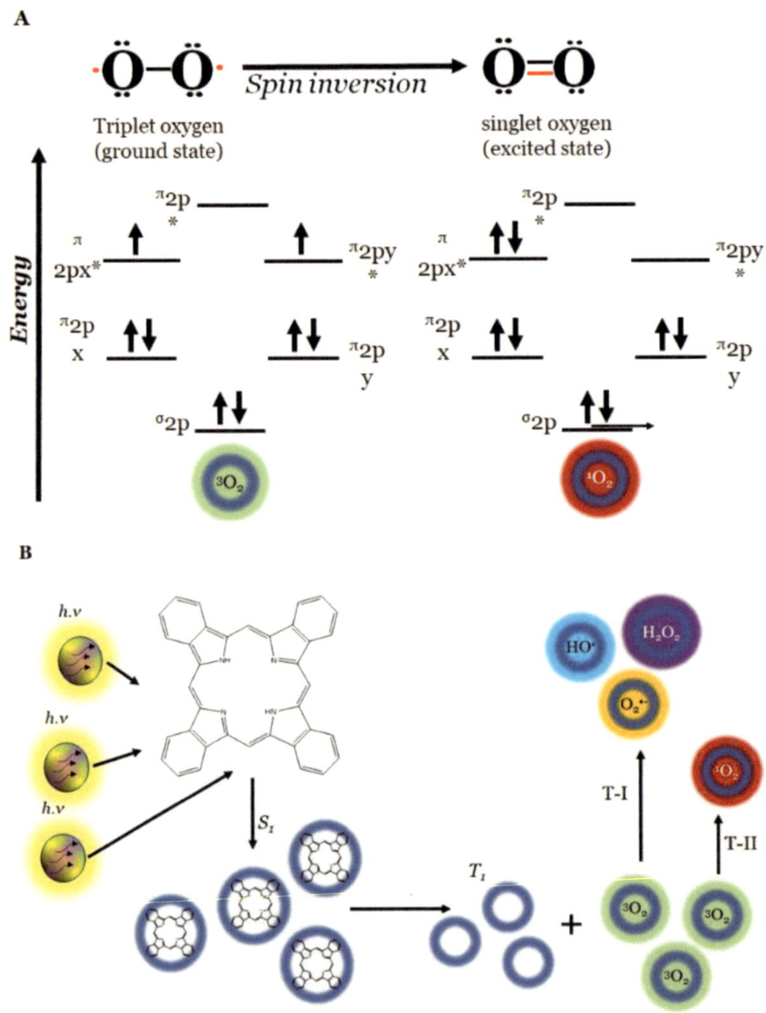

Figure 2.
Basic scheme of the photodynamic reaction. (A) Formation of ROS. (B) Porphyrin group of photosensitizer absorbs a photon that excites it to the short-lived singlet state and may decay by non-radioactive relaxation with heat emission or fluorescence emission to the long-lived triple state. In this triplet state, PS can interact with molecular oxygen in two ways, type-1 and type-2, leading to the formation of oxygen radicals and singlet oxygen [31].

The light is formed by subatomic particles given off by atoms and are endowed with high luminous energy, and energy differences result in different colors called photons. The laser (Light Amplification by Stimulated Emission of Radiation) consists of a monochromatic, non-ionizing and highly concentrated beam of light. Each wave has identical coherence in size and physical shape along its axis, producing a specific form of electromagnetic energy. This wave is characterized by spatial coherence, that is, the beam can be well defined. The intensity and amplitude of the beam follow the curve of the Gaussian beam bell as most of the energy is in the center, with a rapid drop at the edges. There is also a temporal coherence, which means that the emission of the single wavelength has identical oscillations over a period. The final laser beam starts in a collimated form and can be emitted over a long distance in this way. However, bundles emanating from optical fibers generally diverge at the tip. When using lenses, all the beams can be precisely focused, and this monochromatic and coherent beam of light energy can achieve the treatment goal [16, 32, 33].

Photosensitizing agents (PS) consist of molecules in the singlet state in their fundamental state because they have two electrons with opposite spins that allow the transport and transfer of light energy for a chemical reaction, where each PS has unique characteristics for successful activation such as wavelength and creep intensity [34–36].

Most of them are derived from endogenous dyes and are characterized by not being toxic to cells. The molecular structure of most PSs used in PDT is based on a tetrapyrrol skeleton. This type of structure occurs naturally in several important biomolecules, such as heme, chlorophyll, and bacteriochlorophyll, being called "pigments of life" [36]. Therefore, PSs based on porphyrin structures satisfy most of the desirable properties of PSs, such as the high efficiency of singlet generation ($^1O_2$), absorption of the higher wavelengths of the electromagnetic spectrum and a relatively greater affinity for malignant cells, in addition, due to the internal dimensions of the macrocycle cavity and the chelate effect, the porphyrin macrocycle can coordinate transition metals in various oxidation states [36].

4. Bacteriophages and PDT

Resistance to antibiotics spreads rapidly in relation to the discovery of new compounds and their introduction into clinical practice. In addition, the increase in bacterial adaptation can be directly correlated to the scarcity of new classes of antimicrobial agents. In the last decades, synthetic tailoring has been the main strategy to improve the nuclear scaffolding established through analog generation. Although this approach has been beneficial, this research has faced a 'Discovery Void' for 30 years, of which no new class of drugs effective against problematic ESKAPE pathogens. In addition, pathogenic microorganisms generally have the ability to form biofilms. This cellular superstructure may exhibit greater resistance to antibiotics and cause serious and persistent health problems in humans [37–39].

Bacteriophages (phages) are ubiquitous viruses which cause no harm to human or animal cells but are capable to specifically infect, replicate, and kill bacteria [40, 41]. Bacteriophages have been described for delivering successfully antimicrobial agents into bacteria, which consist in a potential alternative for the treatment of infectious diseases [42, 43] caused by bacteria.

The first *in vivo* evidence of effective phage therapy against *Klebsiella pneumoniae*, one of Gram-negative bacterium of ESKAPE group listed in the critical priority tier [44, 45] as serious opportunist in nosocomial infections in the respiratory and urinary tracts, wound sites and blood [46], was demonstrated by Anand et al. [47]. The authors observed significant reduction in the lung lesion severity in the mouse model, suggesting the efficacy of a novel lytic phage VTCFPA43 therapy against virulent *K. pneumoniae* infection by the intranasal route.

According to [48, 49], the delivery systems based on a phage-carrying PS exhibit increased effective killing by the concentrated fluence at the bacterial cell wall, and consequently, reduced side damage to the indigenous microbiota by the site singlet oxygen. Moreover, they investigated the photodynamic effects of the photosensitizer tin (IV) chlorin e6 (SnCe6) (**Figure 3A**) covalently linked to phage 75 on several strains of *S. aureus*, including methicillin- and vancomycin-intermediate strains. Pathogens such as Methicillin-resistant *Staphylococcus aureus* (MRSa) show that antibiotic resistance rates are surpassing 50% in 5 out of 6 world regions of the World Health Organization (WHO) [37, 50]. Results showed that the phage 75 conjugated with SnCe6 was not capable to damage human epithelial cells whereas potently showed bactericide effect against vancomycin-intermediate and MRSa. Additionally, other exogenous photosensitizers (protoporphyrin IX and protoporphyrin diarginate) have been successfully *in vitro* evaluated against clinical strains of MRSa [51].

Acinetobacter baumannii is other important Gram-negative bacterium multidrug-resistant involved in nosocomial infections [51]. Due its capacity to form biofilms, they have the capacity to survive and persist in intensive care unit environment and medical devices [52], what also make of *A. baumannii* one of critical-priority pathogens encompassing the ESKAPE group, for which new antibiotics and combating strategies are urgently needed [11, 12]. In this way, [53] used for the first time the strategy of combining of cationic photosensitizer (NB), structurally modified to produces ROS, and bacteriophages (APB)-based photodynamic antimicrobial agent (APNB) for eradication biofilm formed by multi-drug resistant *A. baumannii*. Both *in vitro* and *in vivo* assays demonstrated that APBN was efficient to treat *A. baumannii* infection, including being more efficient than some antibiotics when evaluated *in vivo*. These results demonstrated the potential of APNB in combating multidrug-resistant bacteria and biofilm ablation.

Candida albicans and more recently *C. auris* are opportunistic polymorphic fungal pathogens, which exhibits almost 40% mortality rates for superficial and systemic infections in humans [54–57]. Likewise, the increasing occurrence of antibiotic-resistant among *C. albicans* strains, demands new approaches to control this life-threatening pathogen [58]. In [59], it reported the photodynamic inactivation of *C. albicans* by the Pheophorbide A (PPA) (**Figure 3B**), a chlorophyll-based

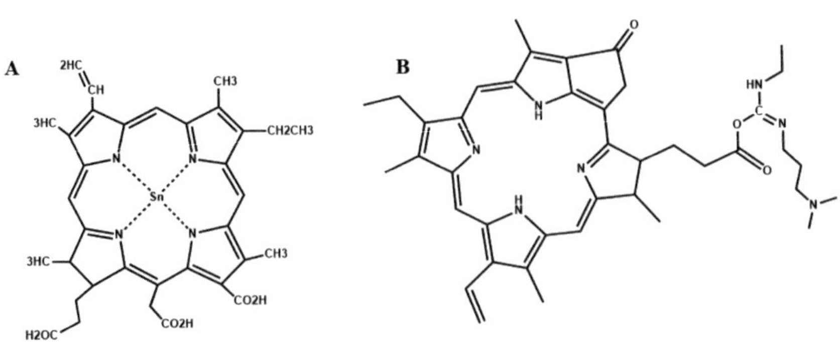

Figure 3.
Chemical structure of (A) SnCe6 and (B) PPA.

photosensitizer crosslinked associated with single-chain variable-fragment phage (JM), which possesses high affinity to β-glucanase mannoprotein (MP65), an essential cell-wall mannoprotein of *C. albicans*. The complex PPA-JM-phage was capable to induce a caspase-dependent apoptosis pathway in *C. albicans.*

The second-generation of PSs exhibits improved photophysical properties in relation to first-generation PSs, which halogens or other substituents are added in the meso- positions of the porphyrin macrocycle. These porphyrins present a better activation of the ring, since the halogens act as removers of electronic density of the ring [18]. Chlorins, which are essentially reduced porphyrins derived from chlorophyll bacteria fetophorides, stable derivatives of chlorophyll varieties that are found in bacteria, are related to porphyrins and are simple to produce. Phthalocyanines and naphthalocyanines, which are derived from azaporphyrin, have high stability, and selectivity [36].

On the other hand, light absorption capacity is an important factor, since most tissues present a comparatively low absorption in the spectral range that extends from 500 nm to about 1500 nm (**Table 1**). This wavelength range is popularly known as the therapeutic window or the diagnostic window (**Figure 4**) [26, 61].

Tissue	**Wavelength (nm)**				
	630	**632.8**	**675**	**780**	**835**
	Optical penetration depth (mm)				
Blood		0.19	0.28	0.42	0.51
Mammary tissue		2.59	2.87	3.12	3.57
Brain (postmortem)		0.92	1.38	2.17	2.52
Brain	1.6				
Lung		0.81	1.09	1.86	2.47

Table 1.
Capacity penetration (mm) of light in different tissues. Adapted from [60].

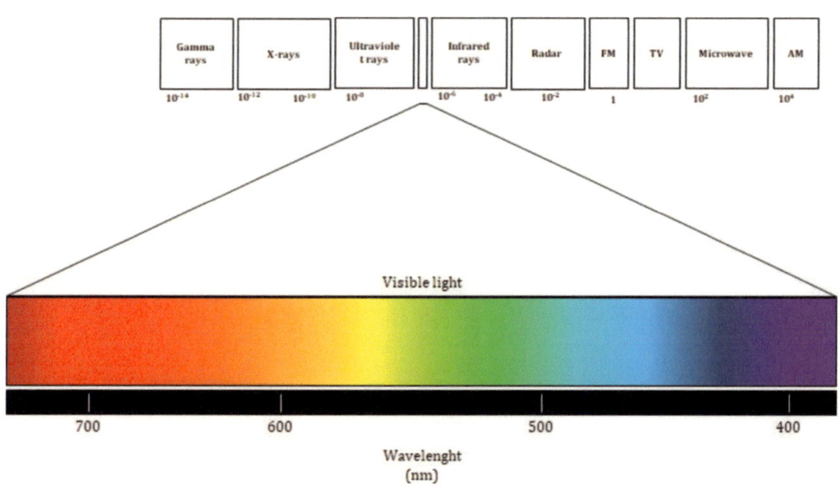

Figure 4.
Electromagnetic spectrum and their respective wavelengths in the region of visible light as a function of the different types of LASERs.

5. Concluding remarks

According to the theoretical bases discussed here, PDT deserves a more central position in the treatment of infectious diseases, since several studies report its enormous potential and applicability on several fronts.

Bacteriophages can act for delivering antimicrobial agents into bacteria, which consist in a potential and efficient alternative for the treatment of infectious diseases.

Delivery systems based on a phage-carrying PS exhibit increased effective killing by the concentrated fluence at the bacterial cell wall, and consequently, reduced side damage to the indigenous microbiota by the site singlet oxygen.

The discovery of new PSs and formulations based on nano structures, in addition to the use in conjunction with already established protocols, PDT shows itself as a strong alternative to conventional treatments.

Conflict of interest

The authors declare no conflict of interest.

Author details

Felipe de Paula Nogueira Cruz[1,2], Andréa Cristina Bogas[1,2] and Cristina Paiva de Sousa[1,2]*

1 Laboratory of Microbiology and Biomolecules – LaMiB, Department of Morphology and Pathology, Federal University of São Carlos, Brazil

2 Biotechnology Graduate Program, Federal University of São Carlos, Brazil

*Address all correspondence to: prokarya@ufscar.br

References

[1] Ortega MÁ, Guzmán Merino A, Fraile-Martínez O, Recio-Ruiz J, Pekarek L, G. Guijarro L, García-Honduvilla N, Álvarez-Mon M, Buján J, García-Gallego S. Dendrimers; dendritic materials: from laboratory to medical practice in infectious diseases. Pharmaceutics. 2020;12, 874. doi.org/10.3390/pharmaceutics12090874.

[2] Prestinace F, Pezzotti P, Pantosti A. Antimicrobial resistance: a global multifaceted phenomenon. Pathogens and global health. 2015;109(7): 309-318. DOI: 10.1179/2047773215Y.0000000030.

[3] Nogueira Cruz FP, Bogas AC, Sousa CP. Plant-Associated microorganisms as a potent bio-factory of active molecules against multiresistant pathogens [Online First], IntechOpen, 2020; DOI: 10.5772/intechopen.93598.

[4] Lewis, K. The cience of antibiotic discovery. Cell. 2020; 181. DOI: 10.1016/j.cell.2020.02.056.

[5] World Health Organization [Internet]. Antimicrobial resistance; 2019 Oct 13 [cited 2020 Dec 20];[Available from: https://www.who.int/news-room/fact-sheets/detail/antimicrobial-resistance. Accessed in 11/23/2020.

[6] Rice LB. Progress and challenges in implementing the research on ESKAPE pathogens. Infection Control and Hospital Epidemiology. 2010;31(Suppl1): S7–S10. DOI: 10.1086/655995.

[7] Founou RC, Founou LL, Essack SY. Clinical and economic impact of antibiotic resistance in developing countries: a systematic review and meta-analysis. PLoS ONE. 2017; 12:e0189621. DOI: 10.1371/journal.pone.0189621.

[8] CDC. Antibiotic Resistance Threats in the United States, 2019. Atlanta, GA: U.S. Department of Health and Human Services. 2019. DOI: http://dx.doi.org/10.15620/cdc:82532.

[9] World Health Organization [Internet]. Global antimicrobial resistance surveillance system (GLASS) report: Early implementation 2020; 2020 May 26 [cited 2020 Dec 19]. Available from: https://www.who.int/glass/resources/publications/early-implementation-report-2020/en/.

[10] Rocha IV, Ferraz PM, Farias TGS, de Oliveira SR. Resistance of bacteria isolated from equipment in an intensive care unit. Acta Paulista de Enfermagem. 2015;28(5):433-439. https://doi.org/10.1590/1982-0194201500073.

[11] World Health Organization. WHO publishes list of bacteria for which new antibiotics are urgently need. 2017 Feb 27 [cited 2020 Dec 19]. Available from: https://www.who.int/news/item/27-02-2017-who-publishes-list-of-bacteria-for-which-new-antibiotics-are-urgently-needed.

[12] Mulani MS, Kamble EE, Kumkar SN, Tawre MS, Pardesi KR. Emerging strategies to combat ESKAPE pathogens in the era of antimicrobial resistance: A review. Frontiers in Microbiology. 2019; 10:539. DOI: 10.3389/fmicb.2019.00539.

[13] Hamblin MR, Ying YH. Introduction: Historical Vignettes from the Field of Photomedicine. Handbook of Photomedicine. CRC Press; 2013. pp. 27-34.

[14] van Straten D, Mashayekhi V, de Bruijn HS, Oliveira S, Robinson DJ. Oncologic photodynamic therapy: basic principles, current clinical status and future directions. Cancers (Basel). 2017; 9(2):19. DOI: 10.3390/cancers9020019.

[15] Warrier A, Mazumder N, Prabhu S, Satyamoorthy K, Murali TS. Photodynamic therapy to control

microbial biofilms. Photodiagnosis and photodynamic therapy. 2020; 3:102090. DOI: 10.1016/j.pdpdt.2020.

[16] Halliday D, Resnick R, Walker J. Fundamentos de Física: Óptica e física moderna. 9nd ed. Rio de Janeiro: LTC, 2012. p. 420.

[17] Keiser G. Fundamentals of light sources. In: Biophotonics. Graduate Texts in Physics. Springer, 2016.pp. 91-118. doi.org/10.1007/978-981-10-0945-7_4.

[18] Setúbal CA. Procura por novos fotossensibilizadores para uso em terapia fotodinâmica. Dissertation, Federal University of Paraná; 2007.

[19] de Broglie, L. XXXV. A tentative theory of light quanta. Philosophical Magazine. 1924;47(278):446-458. doi.org/10.1080/14786442408634378.

[20] Ross EV, Miller L. History and fundamentals of lasers and light sources in photomedicine. In: Hamblin MR, Huang Y, editors. Handbook of Photomedicine. CRC Press; 2013. pp. 35-48.

[21] Gomes ATPC, Neves MGPMS, Cavaleiro JAS. Cancer, photodynamic therapy and porphyrin-type derivatives. Anais da Academia Brasileira de Ciências. 2018; 90(1, Suppl. 2), 993-1026. DOI: 10.1590/0001-3765201820170811.

[22] Allison RR, Bagnato VS, Sibata CH. Future of oncologic photodynamic therapy. Future Oncology. 2010;6(6):929-940. DOI: 10.2217/fon.10.51.

[23] Weishaupt KR, Gomer CJ, Dougherty TJ. Identification of singlet oxygen as the cytotoxic agent in photoinactivation of a murine tumor. Cancer Research. 1976;36(7 PT 1):2326-2329. PMID: 1277137.

[24] Simplicio FI, Maionchi FE, Hioka N. PDT: Aspectos farmacológicos, aplicações e avanços recentes no desenvolvimento de medicamentos. Química Nova. 2002;25(5):801-807. doi.org/10.1590/S0100-40422002000500016.

[25] Benov L. Photodynamic therapy: Current status and future directions. Med Princ Pract. 2015; 24, 1:14-28. DOI: 10.1159/000362416.

[26] Krammer B, Verwanger T. Photodynamic therapy. In: Bergamini G, Silvi S, editors. Applied photochemistry: When light meets molecules. Springer; 2016. pp. 377-396. DOI: 10.1007/978-3-319-31671-0_8.

[27] Chilakamarthi U, Giribabu L. Photodynamic Therapy: Past, Present and Future. Chem Rec. 2017;17(8):775-802. DOI: 10.1002/tcr.201600121.

[28] Lee CN, Hsu R, Chen H, Wong TW. Daylight photodynamic therapy: An update. Molecules. 2020;25(21):5195. DOI: 10.3390/molecules25215195.

[29] Dharmaratne P, Sapugahawatte DN, Wang B, Chan CL, Lau KM, Lau CB, Fung KP, Ng DK, Ip M. Contemporary approaches and future perspectives of antibacterial photodynamic therapy (aPDT) against methicillin-resistant *Staphylococcus aureus* (MRSA): A systematic review. European Journal of Medicinal Chemistry. 2020; 200:112341. DOI: 10.1016/j.ejmech.2020.112341.

[30] Soriano J, Mora-Espí I, Alea-Reyes ME, Pérez-García L, Barrios L, Ibáñez E, Nogués C. Cell Death mechanisms in tumoral and non-tumoral human cell lines triggered by photodynamic treatments: Apoptosis, necrosis and parthanatos. Scientific Reports. 2017;7: 41340. DOI: 10.1038/srep41340.

[31] Ormond AB, Freeman HS. Dye sensitizers for photodynamic therapy. Materials (Basel). 2013; 6(3):817-840. DOI: 10.3390/ma6030817.

[32] Svelto, O. Interaction of radiation with atoms and ions. In: Principles of

Lasers. Springer Science, 2009. pp. 17-79. DOI: 10.1007/978-1-4419-1302-9_2.

[33] Coluzzi, D.; Parker, S.P.A. Lasers in dentistry - current concepts, textbooks in contemporary dentistry. Springer International Publishing, 2017. 1st ed. p. 411. DOI: 10.1107/978-3-319-51944-9_10.

[34] Allison RR, Downie GH, Cuenca R, Hu XH, Childs CJ, Sibata CH. Photosensitizers in clinical PDT. Photodiagnosis photodynamic therapy. 2004;1(1):27-42. DOI: 10.1016/S1572-1000(04)00007-9.

[35] Acedo P, Stockert JC, Cañete M, Villanueva A. Two combined photosensitizers: a goal for more effective photodynamic therapy of cancer. Cell Death Disease. 2014;5(3): e1122. DOI: 10.1038/cddis.2014.77.

[36] Abrahamse H, Hamblin MR. New photosensitizers for photodynamic therapy. Biochemistry Journal. 2016;473(4):347-364. DOI: 10.1042/BJ20150942.

[37] Ling LL, Schneider T, Peoples AJ, Spoering AL, Engels I, Conlon BP, Mueller A, Schäberle TF, Hughes DE, Epstein S, Jones M, Lazarides L, Steadman VA, Cohen DR, Felix CR, Fetterman KA, Millett WP, Nitti AG, Zullo AM, Chen C, Lewis K. A new antibiotic kills pathogens without detectable resistance. Nature. 2015;517(7535):455-459. DOI: 10.1038/nature14098.

[38] Park SR, Tripathi A, Wu J, Schultz PJ, Yim I, McQuade TJ, Yu F, Arevang CJ, Mensah AY, Tamayo-Castillo G, Xi C, Sherman DH. Discovery of cahuitamycins as biofilm inhibitors derived from a convergent biosynthetic pathway. Nature Communications. 2016; 7:10710. DOI: 10.1038/ncomms10710.

[39] Igarashi M. New natural products to meet the antibiotic crisis: a personal journey. The Journal of Antibiotics. 2019;72(12):890-898. DOI: 10.1038/s41429-019-0224-6.

[40] Principi N, Silvestri E, Esposito S. Advantages and limitations of bacteriophages for the Treatment of bacterial infections. Frontiers in Pharmacology. 2019; 10:513. DOI: 10.3389/fphar.2019.00513.

[41] Kasman LM, Porter LD. Bacteriophages. In: StatPearls, editor. Treasure Island (FL): StatPearls Publishing; 2020. PMID: 2963023.

[42] Cahan R. Conjugated and immobilized photosensitizers for combating bacterial infections. Recent Pat Antiinfect Drug Discovery 2013;8(2):121-129. DOI: 10.2174/1574891x113089990010.

[43] Martins WMBS, Toleman MA, Gales AC. Clinical utilization of bacteriophages: a new perspective to combat the antimicrobial resistance in Brazil. The Brazilian Journal of Infectious Disease. 2020;24(3):239-246. DOI: 10.1016/j.bjid.2020.04.010.

[44] Asokan GV, Ramadhan T, Ahmed E, Sanad H. WHO Global Priority Pathogens List: A Bibliometric Analysis of Medline-PubMed for Knowledge Mobilization to Infection Prevention and Control Practices in Bahrain. Oman Medical Journal. 2019;34(3):184-193. DOI:10.5001/omj.2019.37.

[45] Ma Y-X, Wang C-Y, Li Y-Y, Li J, Wan Q-Q, Chen J-H, Tay FR, Ni L-N. Considerations and caveats in combination ESKAPE pathogens against nosocomial infections. Advanced Science. 2019; 7:1901872. DOI: https://doi.org/10.1002/advs.201901872.

[46] Pooi Yin Chung, The emerging problems of *Klebsiella pneumoniae* infections: carbapenem resistance and biofilm formation, FEMS Microbiology Letters. 2016; 363(20)fnw219. DOI: https://doi.org/10.1093/femsle/fnw219.

[47] Anand T, Virmani N, Kumar S, Mohanty AK, Pavulraj S, Bera BCh, Vaid RK, Ahlawat U, Tripathi BN. Phage therapy for treatment of virulent *Klebsiella pneumoniae* infection in a mouse model, Journal of Global Antimicrobial Resistance. 2020; 21:34-41. DOI: https://doi.org/10.1016/j.jgar.2019.09.018.

[48] Embleton ML, Nair SP, Heywood W, Menon DC, Cookson BD, Wilson M. Development of a novel targeting system for lethal photosensitization of antibiotic-resistant strains of *Staphylococcus aureus*. Antimicrob Agents and Chemotherapy. 2005;49(9):3690-3696. DOI: 10.1128/AAC.49.9.3690-3696.2005.

[49] Embleton ML, Nair SP, Cookson BD, Wilson M. Selective lethal photo-sensitization of methicillin-resistant *Staphylococcus aureus* using an IgG-tin (IV) chlorin e6 conjugate. Journal of Antimicrobial Chemotherapy. 2002;50(6):857-864. DOI: 10.1093/jac/dkf209.

[50] Nair DR, Chen J, Monteiro JM, Josten M, Pinho MG, Sahl HG, Wu J, Cheung A. A quinolinol-based small molecule with anti-MRSA activity that targets bacterial membrane and promotes fermentative metabolism. The Journal of Antibiotics. 2017;70(10): 1009-1019. DOI: 10.1038/ja.2017.79.

[51] Grinholc M, Szramka B, Olender K, Graczyk A. Bactericidal effect of photodynamic therapy against methicillin-resistant Staphylococcus aureus strain with the use of various porphyrin photosensitizers. Acta Biochim Pol. 2007;54(3):665-670. Epub 2007 Aug 28. PMID: 17726547.

[52] Vincent JL, Rello J, Marshall J, Silva E, Anzueto A, Martin CD, Moreno R, Lipman, J, Gomersall C, Sakr Y, Reinhart K, EPIC II Group of Investigators. International study of the prevalence and outcomes of infection in intensive care units. JAMA. 2009;302(21):2323-2329. DOI: 10.1001/jama.2009.1754.

[53] Pakharukova N, Tuittila M, Paavilainen S, Malmi H, Parilova O, Teneberg S. Structural basis for *Acinetobacter baumannii* biofilm formation. Proceedings of the National Academy of Science. 2018;115(21):5558-5563. DOI: 10.1073/pnas.1800961115.

[54] Ran B, Yuan Y, Xia W, Li M, Yao Q, Wang Z, Wang L, Li X, Xu Y, Peng, X. A photo-sensitizable phage for multidrug-resistant *Acinetobacter baumannii* therapy and biofilm ablation. Chemical Science. 2020. DOI:10.1039/d0sc04889e.

[55] Gao J, Wang H, Li Z, Wong AH, Wang YZ, Guo Y, Lin X, Zeng G, Liu H, Wang Y, Wang J. *Candida albicans* gains azole resistance by altering sphingolipid composition. Nature Communications. 2018;9(1):4495. DOI: 10.1038/s41467-018-06944-1. Erratum in: Nature Communications. 2019;15;10(1):317.

[56] Ksiezopolska E, Gabaldón T. Evolutionary emergence of drug resistance in *Candida* opportunistic pathogens. Genes. 2018; 9:461. DOI: 10.3390/genes9090461.

[57] Carolus H, Van Dyck K, Van Dijck P. *Candida albicans* and *Staphylococcus* species: A threatening twosome. Frontiers in Microbiology. 2019; 10:2162. DOI: 10.3389/fmicb.2019.02162.

[58] Chen H, Zhou X, Ren B, Cheng L. The regulation of hyphae growth in *Candida albicans*. Virulence. 2020;11(1):337-348. DOI: 10.1080/21505594.2020.1748930.

[59] Dartevelle P, Ehlinger C, Zaet A, Boehler C, Rabineau M, Westermann B, Strub JM, Cianferani S, Haïkel Y, Metz-Boutigue MH, Marban C. D-Cateslytin: a new antifungal agent for the treatment of oral *Candida albicans* associated infections. Scientific

Reports. 2018;8(1):9235. DOI: 10.1038/s41598-018-27417-x.

[60] Usuda J, Kato H, Okunaka T, Furukawa K, Tsutsui H, Yamada K, Suga Y, Honda H, Nagatsuka Y, Ohira T, Tsuboi M, Hirano T. Photodynamic therapy (PDT) for lung cancers. Journal of Thoracic Oncology. 2006;1(5):489-93. doi.org/10.1016/S1556-0864(15)31616-6.

[61] Dong S, Shi H, Zhang X, Chen X, Cao D, Mao C, Gao X, Wang L. Difunctional bacteriophage conjugated with photosensitizers for *Candida albicans*-targeting photodynamic inactivation. International of Journal of Nanomedicine. 2018; 13:2199-2216. DOI: 10.2147/IJN.S156815.

Chapter 5

Role of Phage Therapy in COVID-19 Infection: Future Prospects

Amresh Kumar Singh, Vivek Gaur and Ankur Kumar

Abstract

The pandemic of severe acute respiratory syndrome coronavirus 2 (SARS-CoV-2) was first reported in Wuhan City, China, in 2019. After that, the outbreak has grown into a global pandemic and definite treatment for the disease, termed coronavirus disease 2019 (COVID-19), is currently unavailable. The slow translational progress in the field of research suggests that a large number of studies are urgently required for targeted therapy. In this context, this hypothesis explores the role of bacteriophages on SARS-CoV-2, especially concerning phage therapy (PT). Several studies have confirmed that in addition to their antibacterial abilities, phages also show antiviral properties. It has also been shown that PT is effective for building immunity against viral pathogens by reducing the activation of NF kappa B; additionally, phages produce the antiviral protein phagicin. Phages can also induce antiviral immunity by upregulating expression of defensin 2. Phages may protect eukaryotic cells by competing with viral adsorption and viral penetration of cells, virus mediated cell apoptosis as well as replication. Moreover, by inhibiting activation of NF-κB and ROS production, phages can down regulate excessive inflammatory reactions relevant in clinical course of COVID-19. In this chapter, we hypothesize that the PT may play a therapeutic role in the treatment of COVID-19.

Keywords: phage therapy, phage display, NF-κB, bacteriophage, SARS-CoV-2, COVID-19

1. Introduction

Phages are infections caused by bacterial viruses and they are the most bountiful elements on the Earth [1], due to the introduction of antibiotics their application in clinical practice was immediately overcome in Western countries [2]. Patients with antibiotic-resistant infections are traveling from different spots to Georgia and Poland for phage treatments [3]. Despite all the success cases of patients, phage therapy is still faces significant obstacles, particularly administrative issues. In European countries and United States several on-going efforts are being led for the acceptance of phage therapy [4]. In this chapter, we will first discuss the early and current state of phage therapy, address the major challenges faced by phage therapy treatment in Covid-19 infection and the future prospects in this field [5].

1.1 Early studies of phage therapy (PT)

The efficacy of the phage treatment was confirmed when three patients having the same infection dysentery treated with one dose of the anti dysentery phages and recovered within 24 hours of treatment but his study was not published [1]. However, treatment of infectious diseases of humans reported in 1921 by Richard Bruynoghe and Joseph Maisin, who used bacteriophages to treat staphylococcal skin disease [6]. In addition, d'Herelle used various phage preparations in India to treat thousands of peoples suffering with cholera and bubonic plague [7].

Phages mediate immune regulatory and immunotherapeutic trials that are significant in balancing the immunological homeostasis in human [8]. It was suggested that the viability of PT in autoimmune diseases and it can also be used to for the treatment of infection caused by SARS-CoV-2virus [9]. To determine the infection in mass population, a single sewage test is enough to examine the whole population has been infected or not because RNA of SARS-CoV-2 remains stable with the capsid [10]. However, it has been found from accessible information that although the concentration of the virus in sewage water is high but the transmission risk via this route is very low. This information can play a major role in managing COVID-19 [11].

Phage display technique of producing antibodies was developed for MERS-CoV and effectively applied in light of the fact that bacteriophages have the potential to produce recombinant antibodies (Ab) rapidly [12]. Another Yin-Yang biopanning technique features the chance of utilizing crude antigens for the isolation of monoclonal Ab by phage display method [13]. Before using these expensive techniques, production of artificial Ab was primarily done by using animals but it is a slow process and less cost effective than using bacteriophage display techniques [14]. Bacteriophage could be used to decrease the mortality rate due to Covid-19 pandemic, and for the production of artificial Ab against SARS-CoV-2 in the early stages of infection [15, 16].

2. Interactions between phages and the immune system

It is well known that the immune system plays an important role in phage clearance from animal and human bodies [17]. Components of the mono nuclear phagocyte system (MPS) in the spleen and liver are major sites of phage accumulation. The MPS has been credited for the quick expulsion of administered wild-type phage λ from the human circulatory system [18]. In addition, these phages can directly interact with immune cells by either interacting with cell surface molecules or receptors or through phage transcytosis [19]. Besides the take-up of phages, by Ag presenting cells (APC; e.g., dendritic cells) prompts the activation of B-cells and the exhibition of specific Ab against the phage as shown in **Figure 1** [20].

3. Mode of action

In spite of the huge number of publications on phage therapy, there are only few reports in which the pharmacokinetics of therapeutic phage preparations is depicted [21]. Phages get into the circulatory system of experimental animals (after giving a single dose orally) within 2 to 4 h and they reached into the internal organs within 10 hours and can remains in the human body up to several days [22]. In any case extra exploration is required in order to obtain rigorous

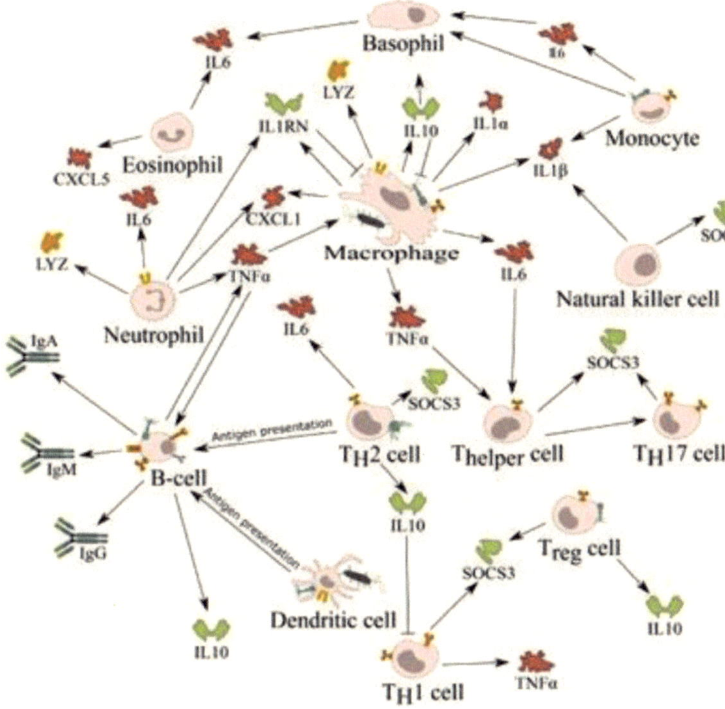

Figure 1.
Interaction of bacteriophage with mammalian immune cells (Belleghem et al. [20]).

pharmacological information concerning lytic phages, including full-scale toxicological research [23]. However, after few years studies reveal that not all phages replicate correspondingly and that there are significant differences in the replication cycles of lytic and lysogenic phages as shown in **Figure 2** [11]. Moreover, it is possible that numerous therapeutic phages act through a common path; however, it may also possible that some therapeutic phages have some distinctive unidentified genes or some unknown mechanisms responsible for lysis of their target bacteria [24]. In a study conducted by *Sulakvelidze et al*. More interpretation of these and common mechanisms is likely to produce

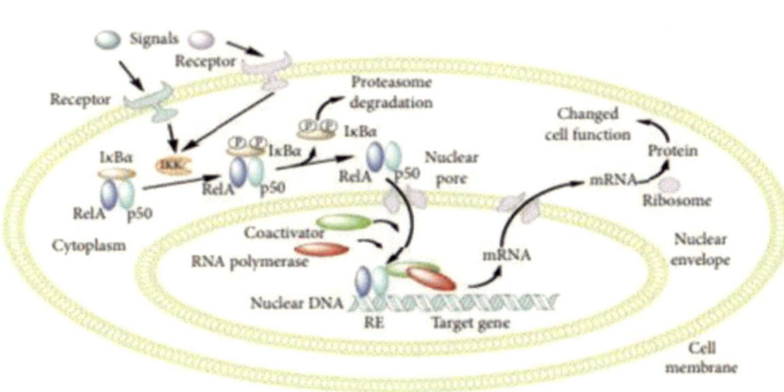

Figure 2.
Mechanism of phage action in bacterial cell (Mishra et al. [11]).

information useful for genetically engineering which was helpful in effective therapeutic phage preparations for the treatment of Coronavirus [7].

3.1 How can ms2 bacteriophage help to fight against coronavirus?

MS2 Bacteriophage is contemplating as a control to study molecular biology processes. It includes viral RNA replication, translation method, and physiology of infected cells. MS2 RNA coding for viral polypeptides includes protein A, coat protein, and RNA replicase complex. The structure of the MS2 virus comprises of Protein A and coat protein makeup.MS2 Bacteriophage can be used as an internal control in RT-PCR testing for COVID-19 to prevent false negative results and to verify the efficacy of the sample preparation and absence of inhibitors in the PCR reaction [25].

4. Phage therapy in humans

Human phage therapy has been practiced in France since 1919, d'Hérelle carried out very extensive studies especially in fowl typhoid and in cholera. In 1921 Bruynoghe and Maisin, Belgium reported that injecting phages targeting *Staphylococcus* near the base of cutaneous boils (furuncles and carbuncles), prompted improvement within 48 hours includes reduction in irritability [26–28].

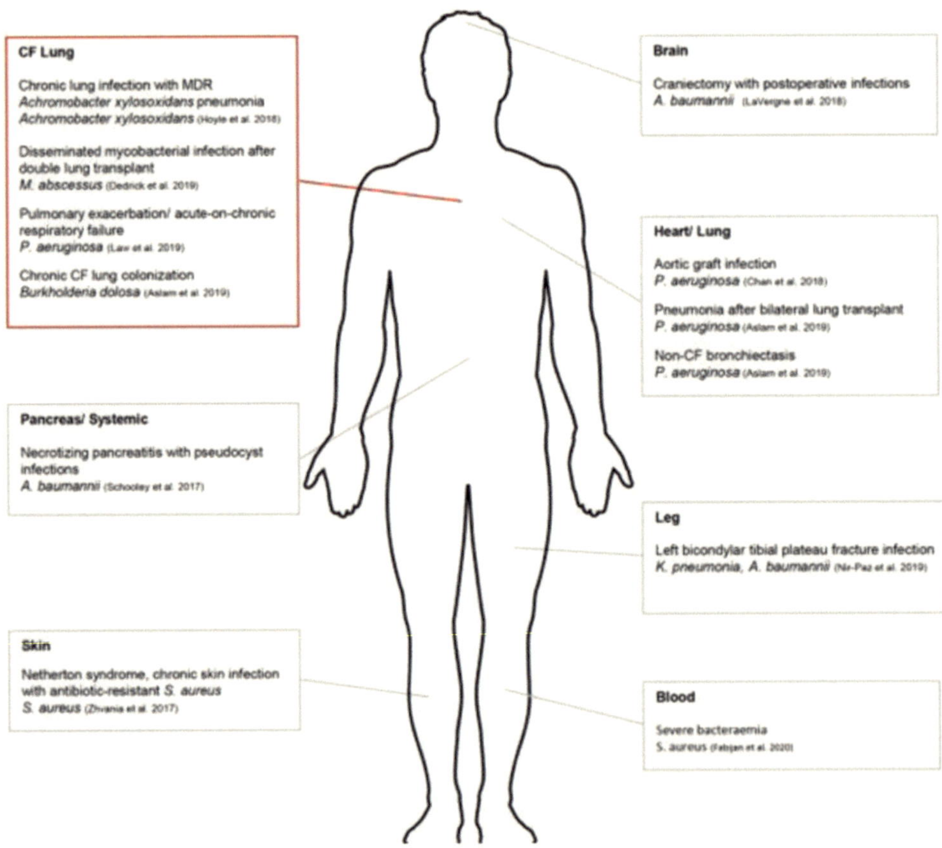

Figure 3.
Schematic diagram indicating areas where phage therapy had been applied clinically (Ng et al. [33]).

A study conducted in by *G. Lang* revealed the utilization of bacteriophage in seven patients with chronic orthopedic infections with antibiotic resistant organisms. He was able to fix two out of seven cases of hip prostheses (after removal of the prostheses) infected by Gram-negative bacteria, one case of tibial osteomyelitis because of the infection caused by*Proteus spp.*, *Staphylococcus aureus* and *Klebsiella spp.*; one instance of septic arthritis of the knee caused due to *Enterobacter* spp. and *Staphylococcus aureus,* one case of septic non-union of the femur due to pan-drug resistant (PDR) *Providencia* [29, 30]. *Henri de Montclos* expressed that phage appear to be safe for human cells though potentially there could be problems associated with their modes of preparation. He also stated about propagation on media produced from animal tissues [31].

The Pasteur Institute stopped making therapeutic cocktails of phages but few French physicians have continued to use phages therapeutically and obtaining their phages from Russia or Georgia. Infections through *Staphylococcus* appear to be the most common target which was treated by phages. In 2011 *Abedon et al.* reported successful phage therapy in two patients from France and Australia who had strong history of antibiotics treatment and other therapies [32]. There are many body places, where phage therapy have been applied and investigated as shown in **Figure 3** [33].

5. A future for phages

The research on phages and their possible antiviral properties are fundamental and should be approved by meticulous *in-vitro* and *in-vivo* studies. If lab research shows some promising results, then it could be possible to have clinical research and randomized stage from one to three human trials to prove their therapeutic utility. Phage therapy may likewise hold promise as a treatment for SARS-CoV-2 [11].

The bacterial growth rate might potentially be diminished by the aerosol use of bacteriophages that prey on the original species of bacteria responsible to cause respiratory failures [34]. This can occur in a self-administrative manner, similar to prey–predator regulation in ecosystem. The remarkable development of the bacteriophage population should allow for a fast clearance, particularly in situations where the bacterial population has already grown significantly [35].

In a study conducted by *Prazak et al.* in 2020, they found the evidences that pneumonia can be treated by nebulized bacteriophages. Target bacteria that commonly cause respiratory problems and selection of bacteriophages can be quickly identified through screening method and by group of experts. Prophylactically administered bacteriophages decreased lung bacterial burdens and improved endurance of antibiotic resistant *S. aureus* infected animals with regards to ventilator-associated pneumonia [36]. It should be ensured to have the right selection of bacteriophages that target both the optimal bacteria and should be most effective against bacterial population growth. The bacteriophages should not interfere with the patient's innate or adaptive immunity. It is also very necessary to rule out that patient does not have antibodies toward bacteriophages used, nor develops any antibodies toward bacteriophages to clear off the bacteriophage earlier than to SARS-CoV-2. If required, quantitative microbiome sequencing can be used potentially in phage therapy [16].

Another obstruction could be a risk of particular species of micro-organism which may develop resistance to the bacteriophage [37]. However, this would be significantly less serious than the drug resistance problem as it would just reduce the efficacy of that one bacteriophage and there is the chance of the bacteriophage also adapting to overcome any resistance to it. They are much specific to one species of bacteria and there is very minor possibility of the bacteriophage damaging any beneficial bacteria but still these things need to be verified through clinical trials. It has to be noted that decrease bacterial growth in critical time of illness allows the patient more time to recover from the SAR-CoV-2 infection [16].

6. Development of a phage display panning strategy

The phage display technology is based on the integration of a gene encoding a peptide or a protein fused with the phage coat proteins was first described by George Smith in 1985 [38]. The most broadly used coat proteins for display are the PVIII and PIII proteins; however, other coat proteins likewise been utilized for display. As a result of its high copy number (~2700 copies), the PVIII protein has been just utilized for the display of small peptides due to conformational issues hampering capsid formation. The PIII system, on the other hand, with its low copy number (5 copies), allows the display of larger molecules such as recombinant antibodies [12]. The first phage display system as shown in **Figure 4** displaying antibodies was explained by *Mc Cafferty et al*. in 1990. They effectively showed variable regions of antibody on phages by using immunoglobulin variable genes of hybridomas and B cells [39]. After its innovation, phage display technology has been extensively used for the research and discovery of antibodies or peptides against a large variety of antigens in many fields of application such as toxicology, drug discovery, immunization, epitope mapping and virus or toxin neutralization by using phage peptide and antibody libraries. Phage display technology has been intensively used for the production of neutralizing antibodies as shown in **Table 1** [40].

Various antibody libraries of different methodologies and strategies have been screened against SARS-CoV-2 spike protein and its receptor binding domain (RBD). Few studies have focused on screening previously developed libraries against SARS-CoV and MERS-CoV and finding cross-reactive antibodies. Others have performed screenings against semisynthetic or synthetic antibody libraries [41]. Phage displayed single-domain antibody was previously developed from llama, which simultaneously neutralizes the S antigen of SARS-CoV and also help in the neutralization of S antigen of the pseudotyped virus SARS-CoV-2 as a bivalent human IgG Fc-fusion protein. A selected antibody has high affinity to RBD, for this a library

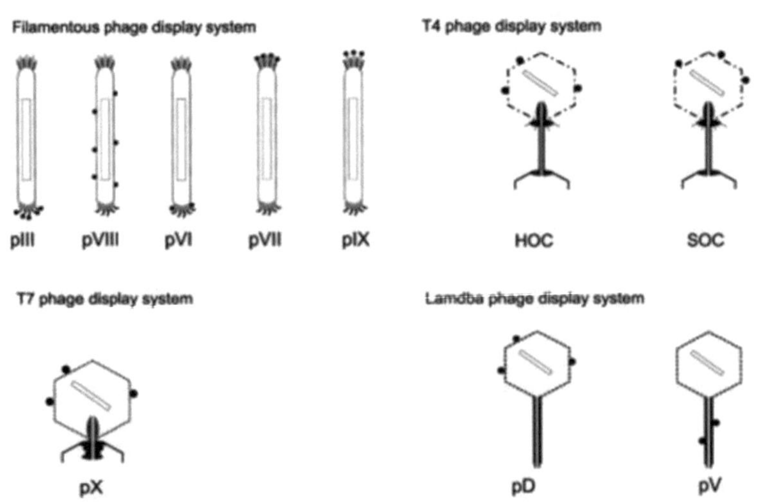

Figure 4.
Schematic presentation of phage display systems (Bazan et al. [12]).

Methods	Original antibody	Reformated antibody	Target region	In vitro/in vivo mode
Phage display	Single-domain antibody from llama	Bivalent human IgG Fc fusion protein	SARS-CoV-2 Spike antigen	Pseudotyped virus and SARS-CoV-2 virus neutralization assay
	Synthetic human Fab library	CDR3 Diversification by mutations	SARS-CoV-2 Spike antigen RBD	Pseudotyped virus neutralization assay
	Single-domain antibody	Grafting naive CDR regions into the framework region of an allele in human antibody heavy chain variable region	RBD domain and the S1 subunit of SARS-CoV-2	Pseudotyped virus neutralization assay
	Naive human scFv antibody	Human IgG1 antibody (4A3)	SARS-CoV-2 RBD	Pseudotyped virus neutralization assay
	Domain library	Fused with human Fc	SARS-CoV-2-RBD	Pseudotyped virus and SARS-CoV-2 virus neutralization assay

Table 1.
*Phage display strategies for neutralizing antibody development (*BalcioĞlu et al. *[15]).*

constructed and screened against the RBD domain of the SARS-CoV-2 spike antigen known as phage displayed synthetic human Fab library [42].

ELISA and pseudo typed virus neutralization assay. A phage-displayed single-domain antibody library has been developed by grafting naive CDR regions into the framework region of an allele in the human antibody heavy chain variable region. They made affinity selection against the RBD domain and the S1 subunit of SARS-CoV-2 and chose several neutralizing antibodies, including a "cryptic" epitope located in the spike's trimeric interface.A site directed screening was performed in a naive human scFv antibody library and domain antibody library by phage display against SARS-CoV-2 RBD. After several rounds of screening, they obtained 9 enriched clones from the domain antibody library and a single clone from the scFv antibody library. The scFv clone was reformatted into a human IgG1 antibody, while the domain antibody clones were fused with human Fc tag. A potential neutralizing effect of these recombinant antibody structures revealed with pseudotyped virus neutralization assay [15].

The future of phage therapy is not necessarily to replace current therapies, rather there is potential for clinical applications to enhance and provide another treatment for infections. Research in this area is likely to grow at an exponential rate. However, the full potential of phage therapy can only be accomplished when there is transparency and an eagerness to share knowledge as well as resources. Preferably, phage libraries should be freely accessible through a network of collaboration, information on preparation and delivery methods for phages implied for clinical usage should be well documented. Phage articulation and delivery are also critical considerations in order to direct activity to targeted areas and maximize efficacy. In fact, use of phage therapy already appears to be as of now gives off an impression of being composed in different nations, and by major public health institutes such as Therapeutic Goods Administration (TGA) (Australia), Food and Drug Authority (FDA) (United States of America) and the European Medicines Agency (EMA) (Europe). Importantly, a universal code of ethics should be established and regulatory bodies reach a consensus on the exchange of information, usage of phages as treatment and reporting of treatment outcomes. Due to the critical nature of the rise of multiple drug resistance (MDR), expanding the urgency for phage therapy to be implemented as standard consideration, alternative therapies to be translated into clinical applications need to be expedited. A concerted effort with both national and international partners could see phage therapy being translated into standard care in the next 5 years [33].

7. Phages as potential inducers of antiviral immunity

There are also data suggesting that phages may drive antiviral immunity by inducing antiviral cytokines, for example, IFN-α and IL-12. An experimental study that phage RNA may induce IFN-α in human granulocytes [43]. Recently, *Sweere et al.* demonstrated that Pf phages (and phage RNA) endocytosed by leukocytes trigger TLR3-dependent pattern recognition receptors and inhibit TNF-driving type I IFN production [44]. The phage-dependent virucidal sign in the lungs could be happen in the phage has capable enough to penetrate the body organ through various routes; therefore phage therapy has been applied successfully in respiratory tract. Intriguingly a fine respiratory microbiome including bacteriophages, during the event of viral pathogens even such as Corona virus is also related with quite low percentage of phages. Recent data indicate that Lactobacillus, *E. coli* and *Bacteroides* phages and phage DNA may stimulate IFN-γ production via TLR9 activation. IFN-γ is another potent antiviral cytokine. Although, the increase TNF level might cause significant risk of virus replication. Hence, a therapeutic agent could regulate TNF production to keep the values at normal level for patient could be appreciated. Pre-clinical studies suggest that viral pneumonia may be cured by anti-TNF therapy. As increased levels of TNF are in blood samples and tissue from patients with COVID-19 may be inhibit TNF production through phage, which is confirmed by other author's previous reports that showed phage may down regulate TNF-α level in serum and lungs of mice with experimental acute pneumonia. Interestingly, clinical phage therapy may reduce TNF production when its pre-treatment level is high and increase it in low responders [45, 46].

These informations might be considered as a relevant argument for phages as a potential agent that could help to decrease TNF levels, allowing for appropriate antiviral immune responses in COVID-19 while reducing the risks of excessive immunosuppression. Different Phages may also interact with TLR [47]. TLR2 is involved in antiviral responses as a result of recognition of the repeating protein subunit patterns common to many viral capsids. [*Induction of Antiviral Immune Response through Recognition of the Repeating Subunit Pattern of Viral Capsid Is Toll-Like Receptor 2 Dependent*]. Other antiviral effects could be mediated by the A5/80 Staphylococcal phage through its ability to increase the expression of the IL-2 gene. IL-2 drives NK cell activity, which is important in defines against viral infections [47]. Phage can also induce antiviral immunity by up regulating expression of defense in IL-2, and recently shown that the T4 phage may induce a marked up regulation of gene coding for hBD2, a multifunctional peptide expressed mainly in epithelial cells with antiviral activity. Virus replication disrupt by the peptide through the binding of the virus by hBD2, decrease viral replication and modulation of signaling pathway essential for virucidal effects, even do the recruitment of immune cells contributing to antiviral activity leading to down regulation of cytopathic effects in human alveolar and laryngeal epithelial cells [48]. In some experiment studies in mice have revealed a co-relation between beta-defensin expression and pulmonary immunity. Moreover, participation of hBD2 in antiviral defenses in the respiratory tract has been confirmed in human disorders [49].

It was advised that phages could be reintroduced for the treatment of not only bacterial, but also other infections such as viral and fungal infections (Adv, *Epstein–Barr virus, Aspergillus fumigatus, Candida albicans*). It showed that there is evidence that proof phage could be comprised in current treatment being studied for re-purposing in the therapeutic treatment of COVID-19. According to *Gorskiet et al.* phage in COVID-19 could be in an adjunct antiviral therapy, which is quite similar to the current trend ofcombined phages with antibacterial treatment in bacterial infections. In other way, a standard phage therapy could be considered for the treatment of bacterial complications of COVID-19, which occur in >40% of patients [45, 50]. Phages may act as shield for eukaryotic cells by competing with surface assimilation and viral penetration of cells; virus mediated, programmed cell death as well as viral replica. Phages may also arouse antiviral immunity during contributing to a equal immune response. Moreover, by inhibiting activation of NF-κB and ROS production, phages can down regulate extreme inflammatory reactions relevant in pathology and clinical course of COVID-19. The data presented in this which was judged are often preliminary but suggest that further studies centered on the potential of phage therapy as at least an adjunct treatment of COVID-19 are warranted. Both general and remote safety of phage therapy was corroborating in human viral diseases. Therefore, extensive studies comprising relevant clinical trials are needed to prioritize applicability of phage to help fighting against COVID-19 pandemic [45].

8. Production of industrial phage propagation strains

The development of new page-based resources using traditional methods can be an on-going issue that may require hundreds of species to be treated with plasmids, active prophages, perhaps other mobile genomic elements. However, given the recent breakthroughs in synthetic biology and advances in re-integration with genetic engineering methods, this need not be the case. Even a given phage infects a particular type of bacterium strain from the affected species depending on the bacterial characteristics and the phage [45].

Metabolic compatibility of a bacterium with a phage to support the propagation of the phage in an already existing infection appears to be specific to certain species, but is sometimes extended to more than one species of bacteria of the same or different genera. Definitions of phage acquisition differentiation encoded by a variety of similar species include genes that include phage receptors or their means of integration and restriction-modification systems associated with the phage. In addition, bacteria, encrypt phage defense process but these mechanisms fortify the bacterium itself from infection through certain pathogens or through the propagation of phage, or induce apoptosis to protect people from the spread of the disease [45].

The distinct indicators of phage determinants are reversible between the strains of given species. Bacteria can gain or lose sensitivity to an appropriate phage or the ability to support the phage development by mutation-recombination-, or horizontal genetically modified changes in their phage orientation or phage defense determinants [51]. There are so many genes which are analogous with phage resistance or susceptibility exposure is carried by mobile genetic elements. Key features of Phage that are important of a metabolically-compatible host include the interaction of phage receptor-binding proteins and receptors on the surface of the bacterial cell, alignment of the phage genome with the bacterial restriction-modification mutation system, or the ability to prevent bacterial action by bacterial restriction-modification systems or by encoding efficient anti-restriction mechanisms. In addition, to infect bacteria, phage reproduce effectively, protein-induced phage allows them to overcome bacterial phage-resistant strains, such as anti-CRISPR proteins and proteins that inhibit the action of Abi systems or toxin-antitoxin (TA) [51, 52].

The structure of each phage and its infectivity for a specific host are determined by the genetic makeup of the phage. The only factors determined by phage handling are considered to be some epigenetic alterations, which are patterns related to host DNA methylation [52]. They have a significant impact on the functioning of new host infections by a phage; play a very important role in horizontal gene transfer through bacteriophages. Therefore, in addition to the species-specific metabolic pathways specific to supporting the efficient propagation of a given phage and which should be equipped with surface receptors for this phage attachment, envelope structures of cell susceptible to the action of the phage lytic protein, and a block-conversion modification system that will allow in this case to infect the desired set of phage in clinics [45, 53].

Removal of such strains of genetic determinants of other phage defense mechanisms (e.g., CRISPR/Cas, Abi, or TA loci), if there is a genetic mutation, can extend the number of phages it can propagate to its cells to phage infecting the strains of the same species and uses the same receptors, but is in capable to overcome the suitable defense. The discovery of sensitivity to several specific phages upon the abolishment of various bacterial phage defense systems has been demonstrated in a number of cases. A good future strategy for finding the therapeutic phage propagation strains of desired properties may be the construction of a bacterial chassis of selected clinically relevant pathogenic species. In synthetic biology, the chassis refers to the microorganism that serves as the basis for genetic engineering and to support them by providing resources for basic tasks, such as replication, transcription, and translation mechanisms [45, 53–55].

The common strains of bacterial chassis that will serve as the basic platforms for construction of industrial phage propagation should have genomes reduce their complexity and unnecessary genetic content by the depletion of most of the transposable element as well as virulence and phage resistance determinants method called as a top-down strategy of the genome reduction process [54]. In addition, they ought to be prepared for the introduction or exchange of genomic modules which enable these strains to function as microbial cells in the use of selected treatment phage. Methods to allow the elimination of mobile genetic elements and other genes are used for genetic reshuffling recombineering, oligo-mediated allelic replacement, or genome editing using CRISPR/Cas-assisted selection of clones for model bacteria, or even on a genomic-wide scale. A repertoire of engineering tools that enhances genomic deceptive ability in bacteria other than *E. coli* uses new and ever-evolving techniques, providing ways to classify genomes belong to particular genera represented by problematic bacterial pathogens, including potential phage propagation strain [45].

The results of studies on micro-organisms that were cured of some or most of the recombinogenic or mobile genetic elements (including prophages) indicate many more benefits. The strain, *Escherichia coli* K-12 with a genomic reduction by approximately 15% by the removal of mobile DNA and cryptic virulence genes. Due to these changes this strain preserved good growth profile and protein production as well as the accurate propagation of recombinant genes and plasmids that could not be stably propagated in other strains [56]. Apart from phage capacity in combating different bacterial infections, emerging evidence suggests role of phages in viral infections as well treatment. Many viral illnesses do not have specific treatment and same antiviral drugs have been used for different viral diseases [57]. Thus, in our opinion, the construction for the propagation of therapeutic phages, of chass is strains equipped with certain phage susceptibility determinants and depleted of phage resistance determinants as well as certain mobiles genetic elements or virulence determinants will not only ensure the safety of therapeutic phage preparations, but will also reduce the cost of phage production substantially [58].

This reduction will be a result of:

i. It helps in reducing the number of strains required for the production of different types of phages.

ii. No need of evaluating phage preparations for the composition (temperate phages and toxins) of undesired elements.

iii. It helps to increase the fitness and stability of these strains in the commercially production of therapeutic phages.

In addition, to single fundamental strain establish for a microbial species can serve as a platform for the enrichment of its genome with several gene cassettes required for the propagation of several phages. Further work to remove additional undesired genomic elements from the genomes of these strains is in progress [45, 59].

9. Phage-based vaccines

Phage-based vaccines offer significant potential advantages by building up a stage approach with the ability to quickly switch the vaccine in response to mutations in the Coronavirus. Also, vaccines based on phages are self-fulfillment, which means they automatically activate and boost immune response, with the ability to show multiple antigens. The therapeutic use of phage in humans is well understood and has a favorable safety measures. Recent studies recommended the immune response to SARS-CoV2 could be transient and need frequent booster vaccinations to manage defensive levels of immunity [60, 61].

The possible advantages of phage-based vaccines incorporate flexibility for route of administration (mucosal and intramuscular), including a potential oral drop, adaptability to virus mutations, quick progression and cost-effectiveness. These advantages, along with the known safety profile, offer hope as a potential tool in reducing this pandemic. Moreover, countries have the potential to increase productivity rapidly. Researchers are fully committed to combat the impact of this public health crisis [60].

9.1 Benefits of phage-based vaccine over other vaccine technologies

- It offers the phages with excellent safety measurements.
- Quickly adaptation of new vaccines to potential mutations in coronavirus.
- Lower cost of manufacturing in comparison with alternative vaccine approaches.
- Self-adjuvanting to provoke immune response

10. Challenges to build phage libraries

The main challenges in treating phage are:

1. Doctors need to know exactly what type of bacteria is causing the infection.
2. They must have various specific phages target that strain, easily available, in fact from a large phage library that can be tested to find the right phage a cocktail that matches the bacteria.

To address latter problem, many pharmaceutical companies are reluctant to committed resources to improve treatment and therapy. That is because phage treatment is almost 100 years old, making it difficult to patent and raise income to allow for the initial cost of development. Lack of regulatory permission to manage the treatment of the page is problematic. Phage cocktails need to be customized for each patient's infection and regularly organized as the bacteria mutate and improve resistance as shown in **Figure 5**. Regulatory agencies such as the US Food and Drug Administration (FDA) currently do not have the necessary review and approval mechanisms to be able to accept your identity and adapt to a greater degree. The experimental design that benefits from genomic sequence and mass spectrometry will soon meet the need for rapid and accurate microbial identification. A second barrier to phage treatment, the need for easily accessible therapeutic phage, could ultimately be met to some extent by the U.S Medical Research Centre and different groups around the world are presently building phage libraries as shown in **Table 2** [63].

Looking forward, other technological innovations could help make the phage treatment more specific and help with patent issues. For an example, phages can at last be developed using CRISPR/Cas9 genetic engineering strategies to kill only

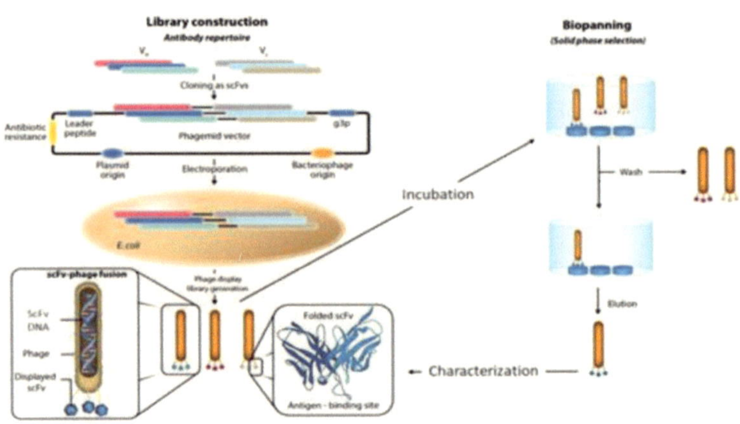

Figure 5.
Phage display method to build library of peptides and proteins variants (source: Almagro et al. [62]).

Library Name	Company/ Laboratory	Repertoire	Display Format	Size
XFab1	Xoma	Naive	Fab	3.1×10^{11}
XscFv2	Xoma	Naive	scFv	3.6×10^{11}
HAL9/10	TU-IB (b)	Naive	scFv	1.5×10^{10}
KNU-Fab	KNU (c)	Naive	Fab	3.0×10^{10}
pIX V3.0	Janssen Bio	Synthetic	Fab	3.0×10^{10}
HuCAL PLATINUM	MorphoSys	Synthetic	Fab	4.5×10^{10}
Ylanthia	MorphoSys	Synthetic	Fab	1.3×10^{11}
PHILO Diamond	ETH Zurich	Synthetic	scFv	4.1×10^{10}
ALTHEA Gold Libraries	GlobalBio/ADL	Semisynthetic	scFv	2.1×10^{10}

Source: Almagro et al. *[62].*

Table 2.
Example of Phage display antibody libraries.

resistant micro-organism. Some agencies there may also be eligible for patents on separate phage or phage cocktails, making them a viable commercial investment [63].

No matter what the future holds for the treatment of the phages, most experts agree that the phage treatment will never completely replace antibiotics. Instead, this method can be used in combination with antibiotics, or as a last resort to protect patients with diseases that have not responded to other treatments. Given the alarming increase in the number of life-threatening multidrug- resistant diseases in recent years, the need to investigate the potential role of phage and other alternative to antibiotic treatments is urgently required [64, 65].

11. Conclusion

The progressing SARS-CoV-2 related COVID-19 pandemic is persistently emerging worldwide and signifying the greatest spotlight on public health, education, travels, and monetary conditions in the current world. As irresistible situations have no borders because there is no single specific therapy that may give effective responses toward COVID-19. Thus, a worldwide activity intends to make phage therapy worldwide overall accessible is required. This obviously requires an active joint effort between countries for overcoming logistic and administrative challenges and among clinicians and researchers for filling current knowledge gap and encouraging advances in the field.

How would it be advisable for us to deal with the current infection prevention and control a strategy which also works after the epidemic? How could we react to similar contagious diseases in the future? These are open questions which require further discussion and research. While phages may have the potential to play a role in the current pandemic, it is also very important to understand that there is no magic stick for this pandemic. The current situation highlights the urgency for adhering to clinical pharmacology, therapeutic, preventative and diagnostics interventions to optimize COVID-19 therapies. The instant and cell free production of synthetic phages, whether designed or not? This had considerable advantages over classically produced natural phages. Implementation of the right patient, right drug, right dosage, and right timing approach helps to reduce the rate of infection. Finally, adaptive designs for COVID-19 will lead to the development of more vigorous infectious disease research infrastructure and funding to help mitigate future pandemics.

Acknowledgements

We would like to thanks Microbiologists and Junior Residents from the Department of Microbiology, Baba Raghav Das Medical College Gorakhpur, UP for providing support in the preparation of this article. We would like to thanks to our laboratory technicians (Mr. Umesh Chaudhary, Mr. Akhilanand Rai, Mr. Jagmohan Prasad) for their support.

Authors' contributions

This work was carried out in collaboration among all authors. Scientific data collection was performed by authors AKS, VG, and AK. The first draft of the manuscript was written by author VG; data was provided and corrected by authors AKS, and AK. All authors commented on previous versions of the manuscript. All authors read and approved the final manuscript.

Author details

Amresh Kumar Singh*, Vivek Gaur and Ankur Kumar
Department of Microbiology, Baba Raghav Das Medical College,
Gorakhpur, Uttar Pradesh, India

*Address all correspondence to: amresh.sgpgi@gmail.com

References

[1] Wilkinson L. Félix d'Herelle and the origins of molecular biology. Medical History. 2001;45(2):294-295.

[2] Britannica. T. Editors of Encyclopaedia. Bacteriophage. Encyclopedia Britannica. 2018. https://www.britannica.com/science/bacteriophage

[3] Rohde C, Wittmann J, Kutter E. Bacteriophages: A Therapy Concept against Multi-Drug-Resistant Bacteria. Surgical infections. 2018;*19*(8):737-744.

[4] Furfaro LL, Payne MS, Chang BJ. Bacteriophage Therapy: Clinical Trials and Regulatory Hurdles. Frontiers in cellular and infection microbiology. 2018;*8*:376.

[5] Pires DP, Costa AR, Pinto G, Meneses L, Azeredo J. Current challenges and future opportunities of phage therapy. FEMS microbiology reviews. 2020;*44*(6):684-700.

[6] Kaźmierczak Z, Górski A, Dąbrowska K. Facing antibiotic resistance: Staphylococcus aureus phages as a medical tool. Viruses. 2014;*6*(7):2551-2570.

[7] Sulakvelidze A, Alavidze Z, Morris JG Jr. Bacteriophage therapy. Antimicrob Agents Chemother. 2001;45(3):649-659.

[8] Gorski A, Dabrowska K, Miedzybrodzki R, Weber-Dąbrowska B, Łusiak-Szelachowska M, Jończyk-Matysiak E, et al. Phages and immunomodulation. Future Microbiology. 017;12(10):905-914.

[9] Fernandes S, São-José C. Enzymes and Mechanisms Employed by Tailed Bacteriophages to Breach the Bacterial Cell Barriers. Viruses. 2018;*10*(8):396.

[10] Michael-Kordatou I, Karaolia P, Fatta-Kassinos D. Sewage analysis as a tool for the COVID-19 pandemic response and management: the urgent need for optimised protocols for SARS-CoV-2 detection and quantification. J Environ Chem Eng. 2020;8(5): 104306.

[11] Mishra VN, Kumari N, Pathak A, Chaturvedi RK, Gupta AK, Chaurasia RN. Possible Role for Bacteriophages in the Treatment of SARS-CoV-2 Infection. International journal of microbiology. *2020*;8844963.

[12] Bazan J, Całkosiński I, Gamian A. Phage display--a powerful technique for immunotherapy: 1. Introduction and potential of therapeutic applications. Human vaccines & immunotherapeutics. 2012;*8*(12):1817-1828.

[13] Lim CC, Woo PCY, Lim TS. Development of a Phage Display Panning Strategy Utilizing Crude Antigens: Isolation of MERS-CoV Nucleoprotein human antibodies. Sci Rep 9, 2019;6088.

[14] Hentrich C, Ylera F, Frisch C, Haaf TA, Knappik A. Monoclonal antibody generation by phage display. In: Vashist SK, Luony JHT; eds. Handbook of Immunoassay Technologies. Cambridge, MA: Elsevier. 2018; 47-80.

[15] BalcioĞlu BK, DenİzcİÖncÜ M, ÖztÜrk HÜ, YÜcel F, Kaya F, Serhatli M, et al. SARS-CoV-2 neutralizing antibody development strategies. Turkish journal of biology = Turk biyolojidergisi. 2020;*44*(3): 203-214.

[16] Wojewodzic MW. Bacteriophages Could Be a Potential Game Changer in the Trajectory of Coronavirus Disease (COVID-19). 2020;1(2):60-65.

[17] Jończyk-Matysiak E, Weber-Dąbrowska B, Owczarek B, MiędzybrodzkiR,Łusiak-SzelachowskaM,

Łodej N, et al. Phage-Phagocyte Interactions and Their Implications for Phage Application as Therapeutics. Viruses. 2017;*9*(6):150.

[18] Navarro F, Muniesa M. Phages in the Human Body. Frontiers in microbiology. 2017;*8*: 566.

[19] Stone E, Campbell K, Grant I, McAuliffe O. Understanding and Exploiting Phage-Host Interactions. Viruses. 2019;*11*(6):567.

[20] Belleghem JDV, Dąbrowska K, Vaneechoutte M, Barr JJ, Bollyky PL. Interactions between Bacteriophage, Bacteria, and the Mammalian Immune System. Viruses. 2018;*11*(1):10.

[21] Abdelkader K, Gerstmans H, Saafan A, Dishisha T, Briers Y. The Preclinical and Clinical Progress of Bacteriophages and Their Lytic Enzymes: The Parts are Easier than the Whole. Viruses. 2019;*11*(2):96.

[22] Principi N, Silvestri E, Esposito S. Advantages and Limitations of Bacteriophages for the Treatment of Bacterial Infections. Frontiers in pharmacology. 2019;*10*:513.

[23] Prabhurajeshwar C, Desai PP, Waghmare T, Rashmi SB. An overview of bacteriophage therapy over antibiotics; as an alternative for controlling bacterial infections.Int J Pharm Sci& Res. 2020;11(3):993-06.

[24] Mohamed E, Ross RP, Hill C, O'Mahony J, Mcauliffe O, Coffey A. Bacteriophages and Their Derivatives as Biotherapeutic Agents in Disease Prevention and Treatment. Journal of Viruses. 2014;10:1155.

[25] Overview of MS2 Bacteriophage – Definition and its Usage (h-h-c.com)

[26] Summers WC. The strange history of phage therapy. Bacteriophage. 2012;*2*(2):130-133.

[27] Dublanchet A, Bourne S. The epic of phage therapy. The Canadian journal of infectious diseases & medical microbiology = Journal canadien des maladies infectieuses et de la microbiologiemedicale, 2007;*18*(1):15-18.

[28] Criscuolo E, Spadini S, Lamanna J, Ferro M, Burioni R. Bacteriophages and Their Immunological Applications against Infectious Threats. Journal of immunology research. 2017; 3780697.

[29] Lang G, Kehr P, Mathevon H, Clavert JM, Séjourne P, Pointu J. Bacteriophage therapy of septic complications of orthopaedic surgery. Revue de chirurgieorthopediqueetreparatrice de l'appareilmoteur. 1979;*65*(1):33-37.

[30] O'Flaherty S, Ross RP, Meaney W, Fitzgerald GF, Elbreki MF, Coffey A. Potential of the polyvalent anti-Staphylococcus bacteriophage K for control of antibiotic-resistant staphylococci from hospitals. Applied and Environmental Microbiology. 2005;*71*(4):1836-1842.

[31] Montclos H. Les bacteriophages therapeutique: del'emprirismala biologiemoleculaire. Pyrexie 2002;6:77-80.

[32] Abedon ST, Kuhl SJ, Blasdel BG, Kutter EM.Phage treatment of human infections. Bacteriophage. 2011;*1*(2):66-85.

[33] Ng RN, Tai AS, Chang BJ, Stick SM, Kicic A. Overcoming Challenges to Make Bacteriophage Therapy Standard Clinical Treatment Practice for Cystic Fibrosis. Frontiers in Microbiology. 2021;*11*:593988.

[34] Clokie MR, Millard AD, Letarov AV, Heaphy S. Phages in nature. Bacteriophage, 2011;*1*(1): 31-45.

[35] Nabergoj D, Modic P, Podgornik A. Effect of bacterial growth rate on

bacteriophage population growth rate. MicrobiologyOpen. 2018;*7*(2):e00558.

[36] Prazak J, Valente L, Iten M, Grandgirard D, Leib SL, Jakob SM, et al. Nebulized Bacteriophages for Prophylaxis of Experimental Ventilator-Associated Pneumonia Due to Methicillin-Resistant Staphylococcus aureus. Crit Care Med. 2020;48(7):1042-1046.

[37] Oechslin F. Resistance Development to Bacteriophages Occurring during Bacteriophage Therapy. Viruses. 2018;*10*(7):351.

[38] Antonio AM. Phage display technology: applications and innovations. Genetics and Molecular Biology. 2005;*28*(1):1-9.

[39] McCafferty J, Griffiths A, Winter G, *Chiswell* DJ. Phage antibodies: filamentous phage displaying antibody variable domains. Nature. 1990;348:552-554.

[40] Hammers CM, Stanley JR. Antibody phage display: technique and applications. The Journal of Investigative Dermatology. 2014;*134*(2):1-5.

[41] Xiaojie S, Yu L, lei Y, Guang Y, Min Q. Neutralizing antibodies targeting SARS-CoV-2 spike protein. Stem Cell Research. 2021;50:1873-5061.

[42] Yu F, Xiang R, Deng X. Wang L, Yu Z , Tian S, *et al.* Receptor-binding domain-specific human neutralizing monoclonal antibodies against SARS-CoV and SARS-CoV-2. Sig Transduct Target Ther. 2020;5:212.

[43] Taborsky I, Dolník V. Ability of human polymorhonuclear blood cells to produce interferon after induction with phage double-stranded RNA. ActaVirol. 1977;21(6):499-502.

[44] Sweere JM, Van Belleghem JD, Ishak H, Bach MS, Popescu M, Sunkari V, et al. Bacteriophage trigger antiviral immunity and prevent clearance of bacterial infection. Science. 2019;363(6434):eaat9691.

[45] Górski A, Międzybrodzki R, Żaczek M, Borysowski J. Phages in the fight against COVID-19? Future Microbiol. 2020;15:1095-1100.

[46] Chang YT, Chen CL, Lin CF, Lu SL, Cheng MH, Kuo CF, et al. Regulatory role of GSK-3 β on NF- κ B, nitric oxide, and TNF- α in group A streptococcal infection. Mediators Inflamm. 2013;2013:720689.

[47] Jan B, Maciej P, Ryszard M, Barbara O, Andrzej G. The effects of bacteriophages on the expression of genes involved in antimicrobial immunity.PostępyHigieny i MedycynyDoświadczalnej. 2019;73.414-420.

[48] Miguel M, Alejandro G, Dora R, Carlos C. Effect of Human Beta Defensin-2 in Epithelial Cell Lines Infected with Respiratory Viruses. Journal of Bioanalysis and Biomedicine. 2015;7:136-143.

[49] Meade KG, O'Farrelly C. β-Defensins: Farming the Microbiome for Homeostasis and Health. Front Immunol. 2019;9:3072.

[50] Wang L, He W, Yu X, Hu D, Bao M, Liu H, et al. Coronavirus disease 2019 in elderly patients: Characteristics and prognostic factors based on 4-week follow-up. J Infect. 2020;80(6):639-645.

[51] Trotter M, McAuliffe O, Callanan M, Edwards R, Fitzgerald GF, Coffey A, Ross RP. Genome analysis of the obligately lytic bacteriophage 4268 of Lactococcuslactis provides insight into its adaptable nature. Gene. 2006 Jan 17;366(1):189-199. doi: 10.1016/j.gene.2005.09.022

[52] Furi L, Crawford LA, Rangel-Pineros G, Manso AS, DeSte Croix M,

Haigh RD, et al. Methylation Warfare: Interaction of Pneumococcal Bacteriophages with Their Host. J Bacteriol. 2019;201.

[53] Adams BL. The Next Generation of Synthetic Biology Chassis: Moving Synthetic Biology from the Laboratory to the Field. ACS Synth Biol. 2016 Dec 16;5(12):1328-1330. doi: 10.1021/acssynbio.6b00256.

[54] Szathmáry E. Life: in search of the simplest cell. Nature. 2005 Feb 3;433(7025):469-470.

[55] Fineran PC, Blower TR, Foulds IJ, Humphreys DP, Lilley KS, Salmond GP. The phage abortive infection system, ToxIN, functions as a protein-RNA toxin-antitoxin pair. Proc. Natl. Acad. Sci. USA. 2009;106:894-899.

[56] Ghosh D, Roy K, Williamson KE, Srinivasiah S, Wommack KE, Radosevich M. Acyl-homoserine lactones can induce virus production in lysogenic bacteria: An alternative paradigm for prophage induction. Appl. Environ. Microbiol. 2009;75:7142-7152.

[57] Choudhari O, Ojha UC, Rani A, Spalgais S. Role of bacteriophage in COVID 19. Int J of Scie Res.2020;9(6):61-62.

[58] Lwoff A. Lysogeny. Bacteriol. Rev. 1953;17:269-337.

[59] Łobocka M, Hejnowicz MS, Dąbrowski K, Izak D, Gozdek A, Głowacka A, et al. *Staphylococcus aureus* Strains for the Production of Monoclonal Bacteriophage Preparations Deprived of Contamination with Plasmid DNA. WO 2016/030871 A1. U.S. Patent. 2016 Mar 16.

[60] Adaptive Phage Therapeutics. https://www.aphage.com/science/vaccines/

[61] Greg Merril. Adaptive Phage Therapeutics Receives Department of Defense Award for Development of COVID-19 Vaccine. Adaptive Phage Therapeutics - Adaptive Phage Therapeutics Receives Department of Defence Award for Development of COVID-19 Vaccine (reportablenews.com)

[62] Almagro JC, Pedraza-Escalona M, Arrieta HI, Pérez-Tapia SM. Phage Display Libraries for Antibody Therapeutic Discovery and Development. Antibodies (Basel). 2019;8(3):44.

[63] Phage 101. Center for Innovative Phage Applications and TherapeuticsDivision of Infectious Diseases & Global Public Health.https://therapeutics/research/ Pages/ Phage% 20101.aspx

[64] Sulakvelidze A, Alavidze Z, Morris G.Bacteriophage Therapy. Antimicrob Agents Chemother.2001;45(3):649-659.

[65] Wittebole X, De Roock S, Opal SM. A historical overview of bacteriophage therapy as an alternative to antibiotics for the treatment of bacterial pathogens. Virulence. 2014;5(1):226-235.

Chapter 6

Fecal Virome Transplantation

Derek Lin and Henry C. Lin

Abstract

The gut virome consists of a large population of eukaryotic and prokaryotic viruses that have an emerging role in human health and disease. Growing evidence for the importance of the virome includes recent findings on fecal virome transplantation (FVT) that suggest FVT may have therapeutic potential for the resolution of dysbiosis and treatment of dysbiosis-related disorders. Most viruses in the gut virome are bacteriophages (phages), which have a well-established role in regulating bacterial communities across environments. Phages also influence health and disease by interacting directly with the host immune system. The full extent to which gut phages should be considered as both a target and a tool for microbiome modulation remains to be seen. This chapter will explore the current understanding of the gut virome and the therapeutic potential for FVT.

Keywords: dysbiosis, virome, bacteriophage, phage therapy, microbial therapeutics

1. Introduction

While the role of the bacterial community during fecal microbiota transplantation (FMT) has been the focus of extensive investigation, there has been substantially less examination of the viral community. The growing body of research on the viruses of the gut microbiome, referred to, herein, as the gut virome, points to their role as an important regulator of gut homeostasis [1–4]. This occurs through the modification of microbiome structure, composition, and function by gut bacteriophages [5–9], as well as through direct interaction between the enteric virome and the human immune system [4, 10–14]. In line with the gut virome's regulatory role, several recent studies have shown that fecal virome transplantation (FVT), a procedure similar to FMT albeit filtered to exclude intact fecal bacteria, has potential for resolving gut microbiome dysbiosis and restoring a healthy microbiota [15–18]. The full breadth of possibilities for FVT are only now beginning to unfold, but this emerging field of study has produced exciting findings that suggest FVT may be a versatile therapeutic treatment for multiple forms of dysbiosis. Not only has FVT been used effectively for clinical treatment of *Clostridium difficile* infection (CDI), but promising preliminary results suggest FVT has potential for resolving other dysbiosis such as those associated with diabetes and small intestinal bacterial overgrowth. In this chapter we will be reviewing the current state of gut virome research and discussing the clinical potential for FVT.

2. The gut virome

The gut virome consists of a robust and diverse community of eukaryotic and bacterial viruses, with bacterial viruses (herein referred to as bacteriophages,

IntechOpen

or phages) estimated to make up between 90% [19] and 97.7% [20] of the membership of the gut virome. Approximately 10^{14} viruses, comprised of ~1200 virotypes, reside in the gastrointestinal tract at any given time [21], a population that is roughly 10 times that of gut bacteria but comparable in diversity [22, 23]. However, the ratio of phages to bacteria is approximately 1:1 in the infant gut suggesting the population changes over the course of development [24]. Like microbiome composition, virome composition is highly responsive to diet and when individuals are placed on the same diet, their viromes have been found to converge [25]. However, once established, the human gut virome has been shown to have high inter-individual variation [26], sufficient enough for viromes to be distinguishable between related individuals, such as between infants and mothers [27]. Individual viromes are also stable over time and approximately 80% of gut viruses have been shown to persist over a 2.5 year period [28]. At the population level, metagenomic analysis of viromes has demonstrated that there is a core of shared viruses among viromes within a population that can be used to distinguish between other geographically distinct groups [29, 30]. Recent findings by Manrique and colleagues have suggested that there is also a globally distributed set of core phages that are considered to constitute a "healthy gut phageome;" in part, because the prevalence of these phages is significantly decreased in the setting of inflammatory bowel disease (IBD; [31]. Specific phage community compositions and structures are associated with specific gastrointestinal and extraintestinal diseases including colon cancer [32], IBD [14, 33–36], rCDI [16, 37], and diabetes [15, 38].

3. Phages in the gut

As the dominant members of the gut virome, phages have been the focus of studies on the role of the gut virome in health and disease. Phages are ubiquitous viruses that are the most abundant biological entity in the world and can be found anywhere that bacteria can be found. Studying phages in the gut presents a number of difficulties. The first of which is that phages lack a universal marker, such as the 16 s rRNA gene in bacteria. Second, since phages depend on their bacterial hosts for reproduction and only 39% of bacteria in the gut can be cultured [39], many phages that are associated with the other 61% cannot be cultivated. This means that modern phage research largely depends on costly and labor-intensive viral metagenomics, which also presents challenges due to the immense genetic diversity of phages, the lack of a robust virus metagenomic classification, and still nascent use of bioinformatics to evaluate data set generated from viral metagenomic analysis. Much phage research has revolved around the practice of phage therapy, which has been used for over 100 years in some Eastern European Countries to treat single strain bacterial infections. The emergence of antibiotic resistance has led to phages gaining recent attention for their potential as an alternative to antibiotics [40]. In phage therapy, patients are administered solutions of individual phage strains, or multiple strains (i.e. phage cocktail), which are selected through *in vitro* screening for their specificity to the single bacterial agent causing the infection and for their effectiveness in eliminating that one bacterial species. Much of the interest in phage therapy rather than antibiotics is based on the specificity of phages to target a narrow host range, allowing for the targeted elimination of a bacterial pathogen while leaving commensal bacterial members of the microbiome intact, and the ability of phages to self-propagate upon infection of their bacterial host.

In general, there are two types of phages: lytic and temperate. Lytic phages reproduce via the lytic cycle and temperate phages use the lysogenic cycle (**Figure 1**). Conventional phage therapy uses lytic phages because in the lytic

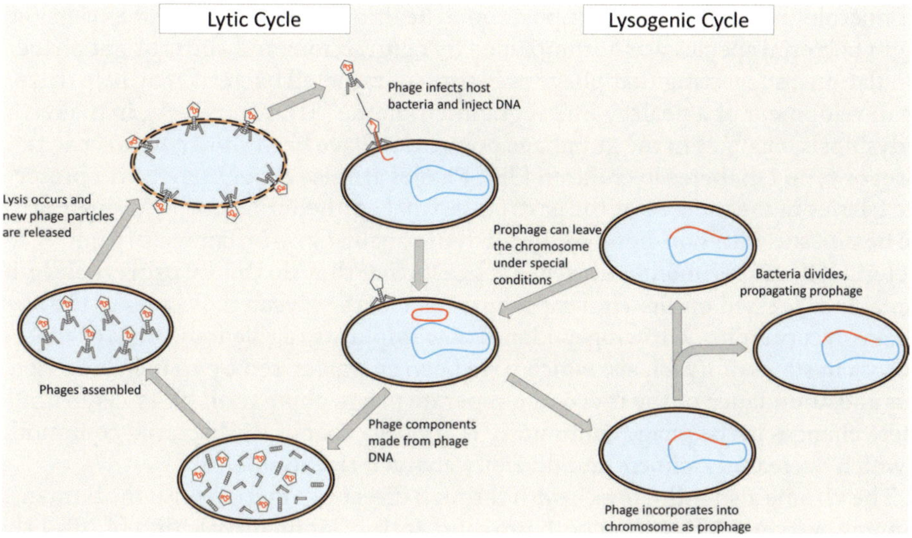

Figure 1.
Reproductive lifecycles of phages.

lifecycle, phages infect a bacterial host, hijack the host machinery for replication of viral progeny, and eventually lyse the host cell and the release of novel phage progeny. In the lysogenic lifecycle, a temperate phage infects a bacterial host and integrates its viral DNA into the bacterial chromosome as a prophage. This process does not always end in cell lysis, instead the prophage can reproduce by propagating with the bacterial chromosome during replication. Harmful environmental stimuli in the gut, such as oxidative stress [41], antibiotics [42], or other unfavorable conditions for the bacterial host [43], can result in the induction of the prophage into the lytic cycle, thereby resulting in the lysis of the bacterial host and release of novel phage progeny. However, Lysogenic (temperate) phages are generally not used in phage therapy because lysogeny is a mechanism for bacteria to exchange DNA so lysogenic phages carry the potential for propagating genes for pathogenesis.

While lytic phages are largely seen as parasitic to their bacterial hosts, temperate phages and their host bacteria have a much more complicated relationship. Temperate phages are important drivers of bacterial evolution [44], in part through their role in horizontal gene transfer between bacterial hosts. Temperate phages are common in the gut and studies have found that a large proportion of bacteria in the microbiome have temperate phages incorporated into their genomes as prophages [21, 45]. For the bacterial host, carrying prophages has several fitness benefits. Prophages encode genes for metabolism, antibacterial resistance, and toxin production (for example, shiga toxin production) [9, 46], thereby conveying functional genes for survival to their bacterial hosts upon integration with the bacterial chromosome. Prophages also protect their hosts from infection by lytic phages through superinfection exclusion [47]. Phage-mediated horizontal gene transfer between bacterial hosts increases rates of genetic recombination and diversification of phage-encoded genes in the gut [48].

Composition, structure, and function of the gut virome contributes to health in a number of ways [49], as reviewed by Mukhopadhya and colleagues [50]. The coevolution between phages and their bacterial hosts is a well-established mechanism for driving the development of microbial communities across environments [44]. This is also the case in the gut environment where phages are thought to modulate the microbiota and, in turn, affect human health. A longitudinal study of gut microbiome and virome composition in healthy infants found that expansion of gut bacterial species was accompanied by contractions and shifts of gut phage populations, suggesting that phage predation of targeted bacteria may help drive the development of a healthy infant gut microbiome [51]. Conversely, in the setting of dysbiosis, changes in the gut phage population have been shown to precede the onset of type 1 diabetes in children [38]. Phages are also thought to form a protective barrier in the mucosa of the gastrointestinal epithelium, thereby providing the host tissue with non-host-derived defense against pro-inflammatory gut bacteria [52]. Experimental evidence suggests that they do this by using their Ig-like domains expressed on the viral capsid to attach to the glycan molecules of the host's mucin glycoproteins. Growing evidence now implicates a role for phages of the mucosa in states of dysbiosis, which have been characterized by an increased richness and abundance of the mucosal temperate phage population [9, 14, 34, 35, 53]. These changes in the phage community is opposite that of the bacterial community in which decreased richness and diversity characterize dysbiosis.

The virome also influences health through direct interaction with the human immune system by triggering both pro- and anti-inflammatory action [4, 10–14]. Phages are capable of activating TLR9-mediated IFNγ, a pro-inflammatory pathway that exacerbates intestinal colitis [14]. Conversely, phages can also ameliorate inflammation through TLR3- and TLR4-mediated interferon-β activation [11]. Several studies have found elevated abundance of phages in the mucosal surfaces of patients with IBD [36, 53]. Other studies have found an expansion of phages from the order *Caudovirales* in the setting of inflammatory bowel disease [34, 54, 55]. Norman and colleagues speculate that phages may contribute to, or be a biomarker for, inflammation and dysbiosis in the gut. Collectively, these studies indicate that phages have an important role in gastrointestinal disorder and potentially, in the corrective response to dysbiosis.

4. Therapeutic potential for FVT

In the setting of FMT, a large population of phages is transferred from the FMT donor to recipient. Feces contain approximately 10^9 virus-like particles per gram, a density similar to that of fecal bacteria, and phages account for upwards of 90% of all fecal virus-like particles [19]. It follows that the large transfer of fecal phages during FMT could have a physiological effect on the FMT recipient. In attempting to examine the role of fecal phages during FMT, several recent studies have not only characterized a state of virome dysbiosis in the setting of recurrent *C. difficile* infection (rCDI), but also have shown that recovery is associated with uptake of a healthy virome from the FMT donor [16, 37, 56]. A study of one FMT patient found that the patient had adopted the donor's phage community after 7 months, even when the patient's microbiome maintained a dysbiotic composition. The microbiome resembled that of the healthy donor a year later [16]. This observation that the adoption of a 'healthy' phage community precedes resolution of dysbiosis may suggest a role for phages in promoting and maintaining a healthy microbiome. This possibility is further substantiated by Zuo and colleagues who found that successful recovery from rCDI after treatment with FMT was associated with a high level of colonization by the donor's phage community in the recipient's enteric virome [37]. Another study showing long-term stability of the FMT recipient's virome found that the donor's phage community maintained colonization of the recipient 12 months after treatment [56]. Similar findings have been observed in clinical trials for FMT as an intervention for pediatric ulcerative colitis [57].

Additional evidence for the active role of phages during FMT comes from studies on fecal virome transplantation (FVT) showing that the sub-bacterial fraction of a FMT (i.e. bacteria removed) can manipulate the composition and structure of a recipient's microbiome [15, 18, 58]. One clinical study found that a fecal suspension that was filtered to remove bacteria, while leaving phages and other sub-bacterial particles intact, was sufficient for effective clinical treatment of rCDI and restoration of a healthy microbiome [58]. Similarly, Kao and colleagues found that a sterilized fecal filtrate was sufficient for treating rCDI [59]. Using another clinically relevant model of dysbiosis, Rasmussen and colleagues demonstrated that a FVT from lean mice was effective at reducing weight and symptoms of diabetes type 2 in obese mice fed a high-fat diet [15]. The investigators also showed that the FVT was able to increase bacterial diversity in the microbiome to the levels in lean mice. The ability for FVT to modulate microbiome composition is further supported by evidence showing that a FVT from high-fat diet-fed obese mice was sufficient for driving microbiome composition of healthy mice towards that of the high-fat diet donor [18]. The investigators also found that a FVT was sufficient for reducing small intestinal bacterial overgrowth (i.e. excess bacterial density in proximal small intestine) in obese mice to the level of healthy controls. In another recent study, investigators found that FVT also prevents necrotizing enterocolitis in preterm piglets [60]. Additionally, there is some speculation that the gut virome has a role in the "super-donor" phenomenon observed during FMT [61]. Collectively, these early studies demonstrate the therapeutic potential for FVT in multiple settings of dysbiosis.

5. Dynamics of FVT-based modulation of the microbiome

The mechanisms through which FVT modifies the recipient's gut microbiome is the subject of ongoing investigation and is likely the result of complex community interactions between donor phages and recipient bacteria, all of which is likely heavily influenced by the host gut environment. Temperate and lytic phages exhibit different population dynamics within microbial communities, and administration of individual strains of exogenous phages into the gastrointestinal tract of mice has been used to study these dynamics [5, 6, 62]. In a gnotobiotic mouse model where the gut is colonized by a defined community of resident gut bacteria, the administration of monocultures of lytic phages exhibiting a narrow host range can reduce populations of their host bacteria through predation [5]. It was also observed that reducing targeted host bacteria subsequently leads to a cascading effect in which populations of non-host bacteria in the microbiome are affected through inter-bacterial interactions. This effect propagated throughout the gut the microbiota with far-reaching consequences for the composition, structure, and function of the microbiota. Additionally, there is some evidence that a phage therapy approach has the potential to control or eliminate bacterial pathogens, such as *Enterococcus faecalis*, in the gut [63]. These studies provide models for studying basic phage-bacterial dynamics in the gut, particularly 'kill-the-winner' population dynamics where lytic phages act as predators leading to a suppression of their bacterial hosts and opening of new ecological niches for non-host bacteria.

Since both temperate and lytic phages are transferred to the recipient during FMT and sustained in the recipient's virome afterwards [17], it is likely that multiple population dynamics are at play in the setting of FVT. In a study using gnotobiotic mice with a defined microbiota, administration of a FVT from human feces resulted in a cascade of changing abundance of different gut bacteria that modeled primarily that of temperate phage-bacteria dynamics [62]. In another study by Bao and colleagues, the investigators found that administration of lytic or temperate phage monocultures into the gut of healthy mice modulated the microbiome by changing relative abundances of host and non-host bacterial populations at both phylum and genus level [6]. Of note, in this particular study, lytic phages promoted a beneficial gut environment while temperate phages promoted conditions that would enable disease to occur. Other co-evolutionary phage-bacteria dynamics that have been observed in microbial communities include 'piggyback-the-winner,' 'arms-race,' and 'kill-the-relative' dynamics, which are reviewed in detail elsewhere [2, 64]. Collectively, these dynamics are thought to contribute to the onset and maintenance of states of dysbiosis in the microbiome and are therefore also likely to have a role in recovery from dysbiosis in the setting of FVT. In the setting of rCDI, it is unclear whether exogenous phages with a broad host range down-regulate *C. difficile* populations or whether they promote a healthier microbiome with less ideal conditions for *C. difficile* colonization.

6. Safety considerations for FVT

While therapeutic application of FMT has been explored in many settings of dysbiosis [65–68], current clinical guidelines recommend that FMT should only be used as a last resort for rCDI due to the various safety concerns [69]. Much of the risk of FMT comes from the transfer of bacteria into an immuno-compromised recipient and the potential of inducing an unanticipated bacteria-driven phenotype (e.g., obesity). Accordingly, FVT may be associated with less risk due to the removal of intact bacteria prior to transplantation. However, since viruses are also capable of eliciting a pro-inflammatory response [12, 14], more research needs to be done to better understand how FVT interacts with the recipient host.

Safe clinical application of FVT will also require a deeper understanding of the viruses that comprise the gut virome. Numerous disease-causing viruses reside in the gut including herpesvirus, papillomaviruses, and hepatitis viruses. Sequencing of the virome has revealed numerous other viruses including bocaviruses, enteroviruses, rotaviruses, and sapoviruses [28]. Many of these viruses have yet to be characterized and their function in the gut is unknown. Given the potential for infection by eukaryotic viruses, a thorough screening of the donor virome must be done to ensure that no harmful eukaryotic viruses are transferred into the recipient. The metagenome of the virome should also be screened since phages can encode genes for virulence factors (e.g., diphtheria toxin, shiga toxin, and botulinum toxin) and antibiotic resistance (e.g., β-lactamases) [70, 71].

Overall, the removal of bacteria is likely to make FVT a safer option than FMT. However, FVT still has safety considerations that must be better understood and effectively taken into account.

7. Conclusion

The emerging field of research focused on the gut virome is still in its infancy, in part due to the difficulty of studying viruses in the gut environment. However, similar to the field of microbiome research, recent work on the gut virome demonstrates how previously overlooked inhabitants of the gut have a profound influence on, and are in fact inseparable from, health and disease. In the setting of FMT, the emerging association between uptake of the donor's phage community and clinical outcome suggests that fecal phages may have an important but not yet fully characterized role in successful treatment of rCDI. Whether FVT will offer a safer or more effective alternative to FMT remains to be seen. We still have yet to determine the full therapeutic potential of FVT, but the promising preliminary findings on FVT suggest it may provide new treatment options for dysbiosis and dysbiosis-associated disorders. Collectively, these recent advances argue for more attention to be given to FVT as a therapeutic tool for microbiome modulation and to the gut virome as a therapeutic target.

Acknowledgements

This work was supported, in part, by the Winkler Bacterial Overgrowth Research Fund.

Conflict of interest

The authors have I. P. rights in related areas.

Author details

Derek Lin[1] and Henry C. Lin[2,3*]

1 Biomedical Research Institute of New Mexico, United States

2 Medicine Service, New Mexico VA Health Care System, Albuquerque, NM, 87108, United States

3 Division of Gastroenterology and Hepatology, University of New Mexico, Albuquerque, NM, 87131, United States

*Address all correspondence to: helin@salud.unm.edu

References

[1] Focà A, Liberto MC, Quirino A, Marascio N, Zicca E, Pavia G. Gut inflammation and immunity: what is the role of the human gut virome? Mediators Inflamm. 2015 Apr 7;2015:326032.

[2] Mirzaei MK, Maurice CF. Ménage à trois in the human gut: interactions between host, bacteria and phages. Nat Rev Microbiol. 2017 Jul;15(7):397-408.

[3] Ogilvie LA, Jones BV. The human gut virome: a multifaceted majority. Front Microbiol. 2015 Sep 11;6:918.

[4] Metzger RN, Krug AB, Eisenächer K. Enteric Virome Sensing—Its Role in Intestinal Homeostasis and Immunity. Viruses. 2018 Mar 23;10(4):146.

[5] Hsu BB, Gibson TE, Yeliseyev V, Liu Q, Bry L, Silver PA, et al. Bacteriophages dynamically modulate the gut microbiota and metabolome [Internet]. bioRxiv. 2018 [cited 2019 Feb 27]. p. 454579. Available from: https://www.biorxiv.org/content/10.1101/454579v1.abstract

[6] Bao H-D, Pang M, Olaniran A, Zhang X-H, Zhang H, Zhou Y, et al. Alterations in the diversity and composition of mice gut microbiota by lytic or temperate gut phage treatment. Appl Microbiol Biotechnol. 2018 Dec;102(23):10219-10230.

[7] Abeles SR, Pride DT. Molecular bases and role of viruses in the human microbiome. J Mol Biol. 2014 Nov 25;426(23):3892-3906.

[8] Moreno-Gallego JL, Chou S-P, Di Rienzi SC, Goodrich JK, Spector TD, Bell JT, et al. Virome Diversity Correlates with Intestinal Microbiome Diversity in Adult Monozygotic Twins. Cell Host Microbe. 2019 Feb 13;25(2):261-72.e5.

[9] Kim M-S, Bae J-W. Spatial disturbances in altered mucosal and luminal gut viromes of diet-induced obese mice. Environ Microbiol. 2016;18(5):1498-1510.

[10] Tetz GV, Ruggles KV, Zhou H, Heguy A, Tsirigos A, Tetz V. Bacteriophages as potential new mammalian pathogens. Sci Rep. 2017 Aug 1;7(1):7043.

[11] Yang J-Y, Kim M-S, Kim E, Cheon JH, Lee Y-S, Kim Y, et al. Enteric Viruses Ameliorate Gut Inflammation via Toll-like Receptor 3 and Toll-like Receptor 7-Mediated Interferon-β Production. Immunity. 2016 Apr 19;44(4):889-900.

[12] Van Belleghem JD, Clement F, Merabishvili M, Lavigne R, Vaneechoutte M. Pro- and anti-inflammatory responses of peripheral blood mononuclear cells induced by Staphylococcus aureus and Pseudomonas aeruginosa phages. Sci Rep. 2017 Aug 14;7(1):8004.

[13] Kernbauer E, Cadwell K. Autophagy, viruses, and intestinal immunity. Curr Opin Gastroenterol. 2014 Nov;30(6):539-546.

[14] Gogokhia L, Buhrke K, Bell R, Hoffman B, Brown DG, Hanke-Gogokhia C, et al. Expansion of Bacteriophages Is Linked to Aggravated Intestinal Inflammation and Colitis. Cell Host Microbe. 2019 Feb 13;25(2):285-99.e8.

[15] Rasmussen TS, Mentzel CMJ, Kot W, Castro-Mejía JL, Zuffa S, Swann JR, et al. Faecal virome transplantation decreases symptoms of type 2 diabetes and obesity in a murine model. Gut. 2020 Dec;69(12):2122-2130.

[16] Broecker F, Russo G, Klumpp J, Moelling K. Stable core virome despite

variable microbiome after fecal transfer. Gut Microbes. 2017 May 4;8(3):214-220.

[17] Draper LA, Ryan FJ, Dalmasso M, Casey PG, McCann A, Velayudhan V, et al. Autochthonous faecal virome transplantation (FVT) reshapes the murine microbiome after antibiotic perturbation [Internet]. Cold Spring Harbor Laboratory. 2019 [cited 2020 Nov 13]. p. 591099. Available from: https://www.biorxiv.org/content/10.1101/591099v1

[18] Lin DM, Koskella B, Ritz NL, Lin D, Carroll-Portillo A, Lin HC. Transplanting Fecal Virus-Like Particles Reduces High-Fat Diet-Induced Small Intestinal Bacterial Overgrowth in Mice. Front Cell Infect Microbiol. 2019 Oct 15;9:348.

[19] Reyes A, Semenkovich NP, Whiteson K, Rohwer F, Gordon JI. Going viral: next-generation sequencing applied to phage populations in the human gut. Nat Rev Microbiol. 2012 Sep;10(9):607-617.

[20] Gregory AC, Zablocki O, Zayed AA, Howell A, Bolduc B, Sullivan MB. The Gut Virome Database Reveals Age-Dependent Patterns of Virome Diversity in the Human Gut. Cell Host Microbe. 2020 Nov 11;28(5):724-40.e8.

[21] Breitbart M, Hewson I, Felts B, Mahaffy JM, Nulton J, Salamon P, et al. Metagenomic analyses of an uncultured viral community from human feces. J Bacteriol. 2003 Oct;185(20):6220-6223.

[22] Lozupone CA, Stombaugh JI, Gordon JI, Jansson JK, Knight R. Diversity, stability and resilience of the human gut microbiota. Nature. 2012 Sep 13;489(7415):220-230.

[23] Sender R, Fuchs S, Milo R. Revised Estimates for the Number of Human and Bacteria Cells in the Body. PLoS Biol. 2016 Aug;14(8):e1002533.

[24] Liang G, Zhao C, Zhang H, Mattei L, Sherrill-Mix S, Bittinger K, et al. The stepwise assembly of the neonatal virome is modulated by breastfeeding. Nature. 2020 May 1;581(7809):470-474.

[25] Minot S, Sinha R, Chen J, Li H, Keilbaugh SA, Wu GD, et al. The human gut virome: inter-individual variation and dynamic response to diet. Genome Res. 2011 Oct;21(10):1616-1625.

[26] Shkoporov AN, Clooney AG, Sutton TDS, Ryan FJ, Daly KM, Nolan JA, et al. The Human Gut Virome Is Highly Diverse, Stable, and Individual Specific. Cell Host Microbe. 2019 Oct 9;26(4):527-41.e5.

[27] Reyes A, Haynes M, Hanson N, Angly FE, Heath AC, Rohwer F, et al. Viruses in the faecal microbiota of monozygotic twins and their mothers. Nature. 2010 Jul 15;466(7304):334-338.

[28] Minot S, Bryson A, Chehoud C, Wu GD, Lewis JD, Bushman FD. Rapid evolution of the human gut virome. Proc Natl Acad Sci U S A. 2013 Jul 23;110(30):12450-12455.

[29] Rampelli S, Turroni S, Schnorr SL, Soverini M, Quercia S, Barone M, et al. Characterization of the human DNA gut virome across populations with different subsistence strategies and geographical origin. Environ Microbiol. 2017 Nov;19(11):4728-4735.

[30] Holtz LR, Cao S, Zhao G, Bauer IK, Denno DM, Klein EJ, et al. Geographic variation in the eukaryotic virome of human diarrhea. Virology. 2014 Nov;468-470:556-564.

[31] Manrique P, Bolduc B, Walk ST, van der Oost J, de Vos WM, Young MJ. Healthy human gut phageome. Proc Natl Acad Sci U S A. 2016 Sep 13;113(37):10400-10405.

[32] Dahiya DK, Renuka. The gut virome: a neglected actor in colon

cancer pathogenesis. Future Microbiol. 2017 Nov;12:1345-1348.

[33] Clooney AG, Sutton TDS, Shkoporov AN, Holohan RK, Daly KM, O'Regan O, et al. Whole-Virome Analysis Sheds Light on Viral Dark Matter in Inflammatory Bowel Disease. Cell Host Microbe. 2019 Dec 11;26(6):764-78.e5.

[34] Norman JM, Handley SA, Baldridge MT, Droit L, Liu CY, Keller BC, et al. Disease-specific alterations in the enteric virome in inflammatory bowel disease. Cell. 2015 Jan 29;160(3):447-460.

[35] Zuo T, Lu X-J, Zhang Y, Cheung CP, Lam S, Zhang F, et al. Gut mucosal virome alterations in ulcerative colitis. Gut [Internet]. 2019 Mar 6; Available from: http://dx.doi.org/10.1136/gutjnl-2018-318131

[36] Duerkop BA, Kleiner M, Paez-Espino D, Zhu W, Bushnell B, Hassell B, et al. Murine colitis reveals a disease-associated bacteriophage community. Nat Microbiol. 2018 Sep;3(9):1023-1031.

[37] Zuo T, Wong SH, Lam K, Lui R, Cheung K, Tang W, et al. Bacteriophage transfer during faecal microbiota transplantation in Clostridium difficile infection is associated with treatment outcome. Gut. 2018 Apr;67(4):634-643.

[38] Zhao G, Vatanen T, Droit L, Park A, Kostic AD, Poon TW, et al. Intestinal virome changes precede autoimmunity in type I diabetes-susceptible children. Proc Natl Acad Sci U S A. 2017 Jul 25;114(30):E6166–E6175.

[39] Browne HP, Forster SC, Anonye BO, Kumar N, Neville BA, Stares MD, et al. Culturing of "unculturable" human microbiota reveals novel taxa and extensive sporulation. Nature. 2016 May 1;533(7604):543-546.

[40] Lin DM, Koskella B, Lin HC. Phage therapy: An alternative to antibiotics in the age of multi-drug resistance. World J Gastrointest Pharmacol Ther. 2017 Aug 6;8(3):162-173.

[41] Diard M, Bakkeren E, Cornuault JK, Moor K, Hausmann A, Sellin ME, et al. Inflammation boosts bacteriophage transfer between Salmonella spp. Science. 2017 Mar 17;355(6330):1211-1215.

[42] Allen HK, Looft T, Bayles DO, Humphrey S, Levine UY, Alt D, et al. Antibiotics in feed induce prophages in swine fecal microbiomes. MBio [Internet]. 2011 Nov 29;2(6). Available from: http://dx.doi.org/10.1128/mBio.00260-11

[43] Oh J-H, Alexander LM, Pan M, Schueler KL, Keller MP, Attie AD, et al. Dietary Fructose and Microbiota-Derived Short-Chain Fatty Acids Promote Bacteriophage Production in the Gut Symbiont Lactobacillus reuteri. Cell Host Microbe. 2019 Feb 13;25(2):273-84.e6.

[44] Koskella B, Brockhurst MA. Bacteria–phage coevolution as a driver of ecological and evolutionary processes in microbial communities. FEMS Microbiol Rev. 2014 Sep 1;38(5):916-931.

[45] Kim M-S, Bae J-W. Lysogeny is prevalent and widely distributed in the murine gut microbiota. ISME J. 2018 Apr;12(4):1127-1141.

[46] Zhang X, McDaniel AD, Wolf LE, Keusch GT, Waldor MK, Acheson DW. Quinolone antibiotics induce Shiga toxin-encoding bacteriophages, toxin production, and death in mice. J Infect Dis. 2000 Feb;181(2):664-670.

[47] Bondy-Denomy J, Qian J, Westra ER, Buckling A, Guttman DS, Davidson AR, et al. Prophages mediate defense against phage infection through diverse mechanisms. ISME J. 2016 Dec;10(12):2854-2866.

[48] Touchon M, Moura de Sousa JA, Rocha EP. Embracing the enemy: the diversification of microbial gene repertoires by phage-mediated horizontal gene transfer. Curr Opin Microbiol. 2017 Aug;38:66-73.

[49] Mills S, Shanahan F, Stanton C, Hill C, Coffey A, Ross RP. Movers and shakers: influence of bacteriophages in shaping the mammalian gut microbiota. Gut Microbes. 2013 Jan;4(1):4-16.

[50] Mukhopadhya I, Segal JP, Carding SR, Hart AL, Hold GL. The gut virome: the "missing link" between gut bacteria and host immunity? Therap Adv Gastroenterol. 2019 Mar 25;12:1756284819836620.

[51] Lim ES, Zhou Y, Zhao G, Bauer IK, Droit L, Ndao IM, et al. Early life dynamics of the human gut virome and bacterial microbiome in infants. Nat Med. 2015 Oct;21(10):1228-1234.

[52] Barr JJ, Auro R, Furlan M, Whiteson KL, Erb ML, Pogliano J, et al. Bacteriophage adhering to mucus provide a non–host-derived immunity. Proc Natl Acad Sci U S A. 2013 May 16;201305923.

[53] Lepage P, Colombet J, Marteau P, Sime-Ngando T, Doré J, Leclerc M. Dysbiosis in inflammatory bowel disease: a role for bacteriophages? Gut. 2008 Mar;57(3):424-425.

[54] Wagner J, Maksimovic J, Farries G, Sim WH, Bishop RF, Cameron DJ, et al. Bacteriophages in gut samples from pediatric Crohn's disease patients: metagenomic analysis using 454 pyrosequencing. Inflamm Bowel Dis. 2013 Jul;19(8):1598-1608.

[55] Fernandes MA, Verstraete SG, Phan TG, Deng X, Stekol E, LaMere B, et al. Enteric Virome and Bacterial Microbiota in Children With Ulcerative Colitis and Crohn Disease. J Pediatr Gastroenterol Nutr. 2019 Jan;68(1):30-36.

[56] Draper LA, Ryan FJ, Smith MK, Jalanka J, Mattila E, Arkkila PA, et al. Long-term colonisation with donor bacteriophages following successful faecal microbial transplantation. Microbiome. 2018 Dec 10;6(1):1-9.

[57] Chehoud C, Dryga A, Hwang Y, Nagy-Szakal D, Hollister EB, Luna RA, et al. Transfer of Viral Communities between Human Individuals during Fecal Microbiota Transplantation. MBio. 2016 Mar 29;7(2):e00322.

[58] Ott SJ, Waetzig GH, Rehman A, Moltzau-Anderson J, Bharti R, Grasis JA, et al. Efficacy of Sterile Fecal Filtrate Transfer for Treating Patients With Clostridium difficile Infection. Gastroenterology. 2017 Mar;152(4):799-811.e7.

[59] Kao DH, Roach B, Walter J, Lobenberg R, Wong K. A51 EFFECT OF LYOPHILIZED STERILE FECAL FILTRATE VS LYOPHILIZED DONOR STOOL ON RECURRENT CLOSTRIDIUM DIFFICILE INFECTION (RCDI): PRELIMINARY RESULTS FROM A RANDOMIZED, DOUBLE-BLIND PILOT STUDY. J Can Assoc Gastroenterol. 2019 Mar 15;2(Supplement_2):101-102.

[60] Brunse A, Deng L, Pan X, Hui Y, Kot W, Nguyen DN, et al. Fecal filtrate transfer protects against necrotizing enterocolitis in preterm pigs [Internet]. Cold Spring Harbor Laboratory. 2020 [cited 2020 Nov 14]. p. 2020.05.25.114751. Available from: https://www.biorxiv.org/content/10.1101/2020.05.25.114751v1.abstract

[61] Wilson BC, Vatanen T, Cutfield WS, O'Sullivan JM. The Super-Donor Phenomenon in Fecal Microbiota Transplantation. Front Cell Infect Microbiol. 2019 Jan 21;9:2.

[62] Reyes A, Wu M, McNulty NP, Rohwer FL, Gordon JI. Gnotobiotic mouse model of phage–bacterial

host dynamics in the human gut. Proc Natl Acad Sci U S A. 2013 Dec 10;110(50):20236-20241.

[63] Bolocan AS, Upadrasta A, Bettio PH de A, Clooney AG, Draper LA, Ross RP, et al. Evaluation of Phage Therapy in the Context of Enterococcus faecalis and Its Associated Diseases. Viruses [Internet]. 2019 Apr 20;11(4). Available from: http://dx.doi.org/10.3390/v11040366

[64] De Paepe M, Leclerc M, Tinsley CR, Petit M-A. Bacteriophages: an underestimated role in human and animal health? Front Cell Infect Microbiol. 2014 Mar 28;4:39.

[65] Vrieze A, Van Nood E, Holleman F, Salojärvi J, Kootte RS, Bartelsman JFWM, et al. Transfer of intestinal microbiota from lean donors increases insulin sensitivity in individuals with metabolic syndrome. Gastroenterology. 2012 Oct;143(4):913-6.e7.

[66] Colman RJ, Rubin DT. Fecal microbiota transplantation as therapy for inflammatory bowel disease: a systematic review and meta-analysis. J Crohns Colitis. 2014 Dec;8(12):1569-1581.

[67] Moayyedi P, Surette MG, Kim PT, Libertucci J, Wolfe M, Onischi C, et al. Fecal Microbiota Transplantation Induces Remission in Patients With Active Ulcerative Colitis in a Randomized Controlled Trial. Gastroenterology. 2015 Jul;149(1): 102-9.e6.

[68] Heath RD, Cockerell C, Mankoo R, Ibdah JA, Tahan V. Fecal microbiota transplantation and its potential therapeutic uses in gastrointestinal disorders. North Clin Istanb. 2018 Feb 12;5(1):79-88.

[69] Mullish BH, Quraishi MN, Segal JP, McCune VL, Baxter M, Marsden GL, et al. The use of faecal microbiota transplant as treatment for recurrent or refractory Clostridium difficile infection and other potential indications: joint British Society of Gastroenterology (BSG) and Healthcare Infection Society (HIS) guidelines. Gut. 2018 Nov;67(11):1920-1941.

[70] Penadés JR, Chen J, Quiles-Puchalt N, Carpena N, Novick RP. Bacteriophage-mediated spread of bacterial virulence genes. Curr Opin Microbiol. 2015 Feb;23:171-178.

[71] Modi SR, Lee HH, Spina CS, Collins JJ. Antibiotic treatment expands the resistance reservoir and ecological network of the phage metagenome. Nature. 2013 Jul 11;499(7457):219-222.

Chapter 7

Tuning Phage for Cartilage Regeneration

Ayariga Joseph Atia, Abugri Daniel Azumah, Bedi Deepa and Derrick Dean

Abstract

The ever-broadening scope of phage research has left behind the simplistic view of studying phages as just model systems in phage biology to a much broader application ranging from ecological management to immunity. Improved throughput technology in crystallography and structural studies has helped our understanding of these systems as supramolecular machines that possess the capacity of self-assembly. The idea of phages as self-assembling supramolecular nano-machines that are bioactive biomaterials in characteristics, tunable and easily producible have lent its utility to recent fields such as regenerative medicine and tissue engineering. Due to low metabolic activity and slow nutrient diffusion within cartilage, damage to this tissue often inevitably consist of slow and delayed regeneration and healing, the restriction of blood from reaching most part of this tissue and the resultant limitations in the availability of oxygen and other essential amino acids dictates a very slow systemic metabolic response also since transports system in this tissue have to employ less speedy forms. Cartilage regeneration therefore is a huge challenge. This chapter takes a look at the application of the phage display technology in cartilage tissue regeneration.

Keywords: self-assembling, supramolecular, bioactive biomaterials, cartilage tissue regeneration, phage display

1. Introduction

In nature, there exist remarkable structural complexities created out of self-assembly, for instance ice crystals from falling snow. In Molecular self-assembly, molecules adopt specific arrangement automatically without the direction of outside source. Phages like liquid crystals behave in such similar fashion, having the ability to self-assembly. Phages are viruses that infect bacterial cells, and also serve as most commercial vectors for recombinant DNA studies. Molecular self-assembly is a key concept in phage chemistry. The components of most phages or viruses in general have an assembly system which usually is directed through non-covalent interactions such as hydrogen bonding, hydrophobic forces, van der Waals forces, and electrostatic etc., leading to the formation of supramolecular assemblies composed of different shapes and sizes [1]. For instance, the interaction of the P22 phage tailspike protein with its capsid to form an infective phage is entirely non-covalent, however, once interaction is complete, bond reversibility is impossible [2]. Molecular self-assembly allows the construction of interesting molecular topologies. This self-assembly system is also crucial in biological systems in the form of the formation of biomolecular condensates in living organisms, also found in oligomerization of protein subunits to form multimers of complex structures [3]. The application of this system therefore is a bottom-up approach, in which components of the phages are directed to self-assembly to achieve a programmed molecular topology, consisting of the desired shape and functional groups.

Most researches have delved into self-assembling filamentous phages, thus shed light on the pathways for their self-assembly. Filamentous bacteriophages such as the *Escherichia coli* K12-infecting Ff phages (F1, Fd or M13) replicate episomally and contain a circular single-stranded DNA packaged into long filaments. These phages are secreted into the environment without lysing their host. The knowledge of phages in general and filamentous phages in particular can played such a vital role in formulating new approaches in fabricating bioactive biomaterials [4] and providing for synergies and opportunities in phage display and tissue engineering approaches.

2. Phage as biotechnological platform for cartilage study, therapy and diagnosis

Due to a low metabolic activity and slow nutrient diffusion within cartilage, damage to this tissue often inevitably consist of slow and delayed regeneration and healing, the restriction of blood from reaching most part of this tissue and the resultant limitations in the availability of oxygen and other essential amino acids dictates a very slow systemic metabolic response since transports system in this tissue have to employ less speedy forms such as transport proteins across the thick ECM. Accidents that cause injury to the knee may sometimes rupture the articular cartilage. Most diseases associated with articular cartilage include the following; 1) osteoarthritis; a condition where the cartilage covering the bones in joints is thinned and sometimes completely worn out. This leads to exposure of the bone ends to friction and erosion which causes bone damage. Aberrant immunometabolism has also been implicated in most phenotypes of osteoarthritis [5]. 2) Rheumatoid arthritis is a chronic systemic autoimmune disease that primarily affects the lining of the synovial joints. This disease is progressive with the pathological mechanism driven via the deterioration of cartilage, bone erosion; hyperplastic synovium and systemic consequences [6]. Most symptoms of rheumatoid arthritis include arthralgia, swelling, redness, joint pain and hence limiting the range of motion [7]. 3) Some other disease/conditions related to cartilage degeneration are relapsing polychondritis [8, 9], achondroplasia [10], costochondritis [11, 12], herniation [13], chondrosarcoma [14], chondroma [15] etc. While biological factors have been well known to play crucial roles in the etiology of these diseases, therapeutic management of these conditions have proved less reliable. Below, we discuss some related works that have been done in the cell-phage research interface, and how the knowledge of both fields could synchronize to help find answers for cartilage regeneration and therapy challenges.

2.1 Using phage to regulate alignment and morphology of chondrocytes

Chondrocytes in articular cartilage have unique alignment with respect to the articular surface, this is crucial for the functional performance of the cartilage. A deeper comprehension of the chondrocytes and collagen alignment is important for a better appreciation of the load bearing and shock absorption function of this tissue. Chondrocytes are organized into four zonal layers in the articular cartilage tissue: superficial zone, middle zone, deep zone and calcified zone. The superficial zone contains elongated and flattened chondrocytes whereas the middle zone has rounded chondrocytes as shown in **Figure 1**. The deep zone and calcified zone have hypertrophic chondrocytes. Cartilage tissue has one of the poorest proliferative capacities and loss of chondrocytes as well as abrasions to the articular surface could give rise to osteoarthritis [16–18]. In cartilage regenerative effort, the interest usually is to produce cartilage with high performance comparable to the natural tissue, and this implies supplying chondrocytes with the right physical and molecular cues to direct their proliferation, differentiation and tissue regeneration. Therefore, guiding the topological and structural organization of the scaffold in which chondrocytes are seeded as well as modulating the molecular cues functionalized to the scaffold is of crucial importance for cartilage tissue regeneration.

In a biomimetic strategy, He and his colleagues synthesized nanofibrous bio-inorganic hybrid materials by using phage as a model biological nanofiber and calcium hydroxyapatite (HAP) as a model inorganic material [19]. They induced the nanofibers self-assembly into phage-cation complex structures through electrostatic interaction between anionic phage nanofibers and the free precursor cations of the inorganic materials. Successful orientation of collagen molecules was also reported. This bioengineered phage bio-nanofibers as biotemplates oriented the nucleation of HAP, formed cluster of structures induced by calcium ions. They observed that the orientations of HAP crystals were formed along Ca^{2+} induced phage bundles, then finally, their co-assembled collagen-phage hybrid bundles induced an aligned nucleation of HAP on them [19].

This gives an excellent platform for cartilage tissue regeneration experiments since collagen fibers arrangement and alignment is a critical measure of cartilage performance. In articular cartilage, the superficial zone consists of mostly type I collagens that are aligned parallel to the articular surface to reduce friction. Hence, the use of such a novel framework for collagen orientation in cartilage regenerative effort will prove useful. The ability to determine the assembly and orientation of collagen or minerals by this co-assembly process in a bio-mimetic scaffold presents

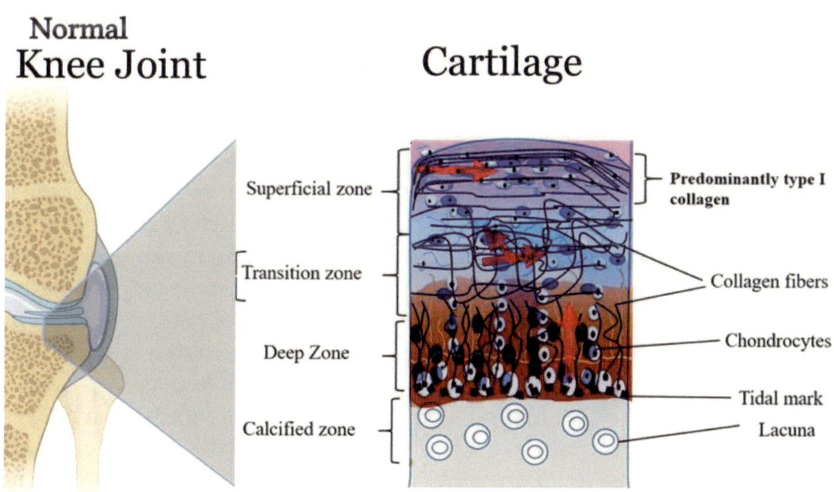

Figure 1.
Cartoon illustration of the anatomical structure of the articular cartilage, depicting the collagen fibers orientation and chondrocytes morphology. The superficial zone consists of flattened and horizontally aligned chondrocytes with also horizontally aligned collagen fibers, predominantly collagen I. the transition zone has rounded chondrocytes, with randomly aligned collagen II fibers; the deep zone has vertically aligned columns of chondrocytes with vertical collagen alignment.

an interesting process for producing excellent cartilage or bone tissue via tinkering the organization and orientation of both proteins and HAP in order to produce superior functional and mechanical properties of cartilage or bone tissue [20].

Young and his colleagues demonstrated that phage-based array chips could be used for an optically readable cell proliferation and morphology assays. They engineered M13 phages that that displayed RGD on its major coat proteins and also functionalized the growth factor, FGF2, on its minor coat proteins. Since M13 can self-assembly, they constructed from them a nanofibrous network scaffold, then grew cells on them. They monitored for biochemical cues displayed by the phage on cell proliferation and morphology. This elegant work allowed for the utility of engineered phages for sensitive monitoring of the effects of functional peptides on cell growth [21].

2.2 Phage used for chondrogenic differentiation of stem cells for cartilage engineering

The chondrogenic potential of mesenchymal stem cells (MSCs) allows for stem cell therapy of damaged cartilage possible. These stem cells can easily be obtained via biopsy from the patient then amplified in the laboratory. This has therefore made MSCs a routinely used cell types for cartilage regeneration [22, 23].

Phage display derived functional peptides has been employed for chondrogenic differentiation of MSCs [24].

TGF-β1- and collagen II-binding peptides were identified through phage display biopanning by Meng and his colleagues [25]. They discovered the peptide HSNGLPL to have high affinity to TGF-β1 receptor, the peptide was then functionalizedbypolyurethane with side propynyl groups via CuAAC click reaction to form nanofiber gel materials with high TGF-β1-binding affinity which acted as an absorbent for TGF-β1 within gels [24, 25]. Their findings demonstrated that their construct induced chondrogenic differentiation of human MSCs *in vitro* and promoted rabbit articular cartilage regeneration.

2.3 Using phage display to initiate cellular signaling

Integrins are transmembrane receptors for extracellular matrix proteins [26]; they play a crucial role in signal transduction in chondrocytes and control cellular attachment, migration, proliferation and apoptosis etc. One of the severally known signaling pathways initiated by integrinsare the Src pathway, this pathway is known to coordinate very vital cellular processes [27], downstream of which include the RHO, SMAD, AKT etc. as depicted in **Figure 2**. While the RHO pathway transduced via integrins acts to regulates actin cytoskeleton [28] leading to cell spread and migration, the AKT through ERK pathway is crucial for chondrocytes growth, proliferation and survival [27, 28]. Interestingly also, downstream of Src, upon ERK pathway activation, the transcription factor SMAD1/5/8 is known to be blocked from nuclear translocation, thereby blocking chondrocyte's hypertrophy, and bone formation processes [29–31]. This step is crucial for forming normal cartilage. While the SMAD signaling pathway will eventually lead to the inhibition of chondrocyte's hypertrophy, the RHO signaling improves cells motility and migration, whereas, the ERK pathway could signal chondrocytic differentiation of MSCs into chondrocytes for cartilage regeneration.

To influence chondrocytic differentiation of MSCs for cartilage regeneration therefore, as diagrammatically illustrated in **Figure 2**, integrin binding peptides such as RGD, IKVAV or DGEA can be genetically engineered to display on the surface of the coat protein of the phage. The displayed peptides which have high

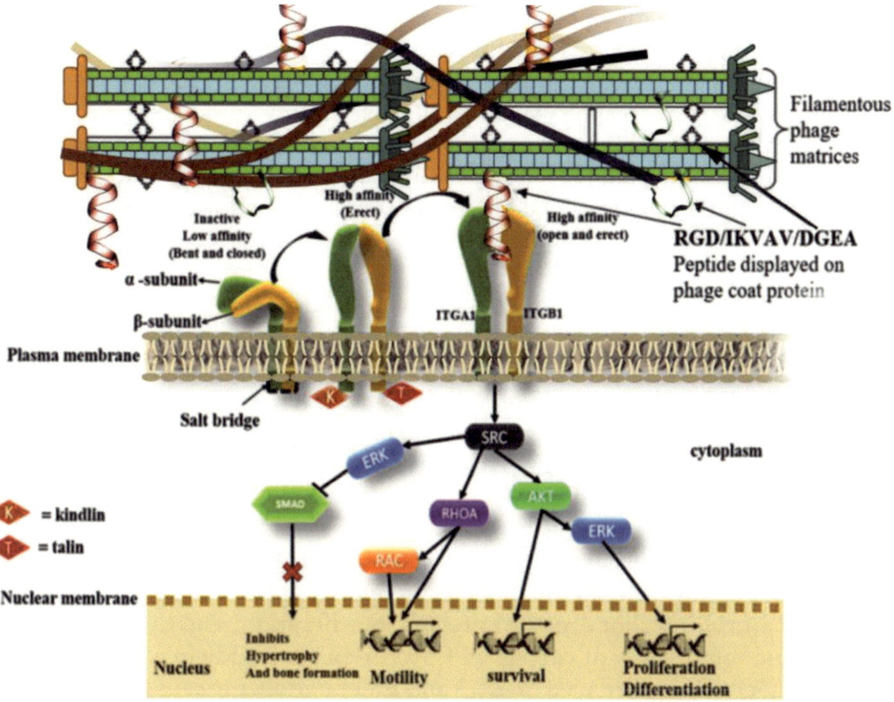

Figure 2.
Diagrammatic illustration of the possible mechanisms of phage displayed peptide on integrin mediated signaling pathway. Integrin binding peptide such as RGD, IKVAV or DGEA can be genetically engineered to display on the surface of the coat protein of the phage. The displayed peptide which have high affinity for integrin receptor will interact with integrin on the plasma membrane, the extracellular subunits of integrin then initiates several conformational changes leading to a more opened and sensitive state from the previously closed and insensitive state, enabling the bound formation between the peptide (ligand) and the receptor. Upon binding to the phage displayed peptide, the integrin receptor activates cascades of signaling including the SMAD, ERK, RHO signaling pathways via signal transduction. While the SMAD signaling pathway will eventually lead to the inhibition of chondrocyte's hypertrophy, the RHO signaling improves cells motility and migration, whereas, the ERK pathway could signal chondrocytic differentiation of prechondrocytes into chondrocytes for cartilage regeneration.

affinity for integrin receptor will interact with integrin on the plasma membrane of the stem cells. This will allow for direct activation of the integrin receptors by the displayed peptide. The process of activating integrin receptor occurs in sequential conformational changes. First, the extracellular subunits of integrin initiate several conformational changes leading to a more opened and sensitive state from the previously closed and insensitive state, the sensitive state allows for bond formation between the peptide and the receptor. Upon binding to the phage displayed peptide, the integrin receptor activates cascades of signaling including the SMAD, AKT, and RHO signaling pathways which could signal chondrocytic differentiation of MSCs into chondrocytes for cartilage regeneration.

2.4 Bio-responsive materials with optimal mechanical and degradation characteristics

The development of scaffolds for cartilage tissue regeneration must include mechanical properties since loading conditions have substantial effect on this tissue. Hence, the optimal mechanical properties of scaffold for cartilage regeneration usually are expected to produce better cartilage tissue formation that must suit the functional role of load bearing. For this reason, it is imperative to carefully regulate the mechanical as well as the degradation properties of the scaffolds used in cartilage engineering. In an ideal case, the biomaterial should eventually be remodeled and replaced by the chondrocytes and the chondrocytes' secreted ECM and studies along these lines have been conducted [32, 33]. The development and application of 'smart' bioresponsive materials that can respond to biological cues or to pathological abnormalities are of great interest to both researchers and clinicians, and this is more so important especially in the case of cartilage tissue regeneration and osteoarthritis therapy in which precise administration of therapeutics with minimal invasiveness is key. A good review on this topic is covered by Yu and his colleagues [33].

In a related study, osteogenic differentiation of mouse preosteoblasts induced by collagen-derived DGEA-peptide on nanofibrous phage tissue matrices was carried out by Yoo and his colleagues [34]. They constructed genetically engineered M13 phage with DGEA-peptide displayed in high density on the major coat proteins and studied the effects of the DGEA-peptides on preosteoblast morphologies. Their results demonstrated that preosteoblasts grown on DGEA-incorporated phage matrices exhibited significant outgrown morphology with early bone cell marker protein expression. In the cartilage tissue, since it is nonvascular, cell to cell communication is slow and most signaling is via ECM embedded proteins, physical cues, peptides, etc. in the tissue extracellular matrices and such play vital role in controlling chondrocyte's growth, proliferation, and ECM molecules deposition and remodeling of cartilage ECM. A replicate study in which the peptide KRTGQYKL is displayed on M13 and prechondogenic cells are grown on such matrices will be of interesting discovery since this peptide is known to induce chondrogenesis [35].

2.5 Assessment of normal cartilage and degenerative cartilage using phage display derived functional peptides

The cartilage is an avascular tissue that expresses high levels of hyaluoronic acid (HA) via the hyaluoronic acid synthase. Classic histochemical analysis of HA are usually performed using Alcian blue or using the HA-specific probe, known as HA-binding protein (HABP), however since HABP is a complex of aggrecans and link proteinsderived from bovine cartilage, published data seems to indicate discrepancies [36]. Zymolik and Mummert via phage display identified a novel HA binding peptide for which they coined pep-1 which demonstrated excellent staining for dermis, however, they recorded sensitivity of pep-1 conformation changes in HA [22]. The pep-1 peptide therefore could serve as HA expression probe for *in situ* detection of hyaluronans since normal cartilage tissue formation is governed by high expression levels of HA, hence assessment of cartilage tissue could be probed via pep-1 too.

The activation of hyaluronan synthase leads to the production and deposition of HA on the ECM of the cartilage for repair and remodeling. It has been shown to modulate inflammation and fibroplasia during wound repair. Tolg et al. using phage display identified another peptide, P15-1 (STMMSRSHKTRSHHV), by biopanning through 7-to 15mer phage display libraries. This 15mer peptide showed similarity to the receptor for hyaluronan mediated motility (RHAMM) binding sequences, and was demonstrated to show high affinity to HA and keenly mimicked the functional properties of RHAMN. In an *in vivo* experiment, P15-1 significantly reduced wound macrophage number, fibroblast number, and blood vessel density compared to negative control peptides in rat wounds and promoted scarless wound healing. They showed that P15-1 blocks RHAMM-regulated focal adhesion kinase pathways in fibroblasts and attenuated fibrotic repair by blocking hyaluronan oligosaccharide signaling [37]. Since the avascular articular cartilage must deal with frictional forces, scar formation is unwanted; therefore the ability to ensure scarless healing is of paramount importance.

Another important molecule of the characteristically thick ECM of cartilage tissue is decorin. It is known to bind to aggrecan to increase its adhesion with other aggrecan molecules and with collagen II fibrils, thereby enhancing the assembly and structural integrity of the aggrecan network in cartilage ECM. At the cellular level, decorin functions to increase the retention of aggrecan in the newly formed matrix of chondrocytes. Also, this molecule increases the adhesion between aggrecan and aggrecan molecules and between aggrecan molecules and collagen II fibrils [38]. It has been shown to inhibit TGF-β and hence prevent tissue fibrosis and promote tissue regeneration. Jarvinen and Ruoslahti genetically displayed a wound-homing CAR peptide (CARSKNKDC) on the decorin surface to form a recombinant CAR-decorin. After intravenous injection of CAR-decorin, these complexes selectively accumulated in the wound sites, and promoted wound healing, without scar formation in a mice wound model [39], this displayed peptide therefore can be employed for cartilage wound healing process, since osteoarthritis is characterized by a persistent deterioration of the cartilage tissue or basically and an non-healing wound.

2.6 Phage used for targeted cartilage tissue drug delivery

By phage biopanning, Pi and his colleagues discovered the chondrocyte-homing peptide, DWRVIIPPRPSA (CAP). They also chemically conjugated the peptide with polyethyleneimine (PEI) to construct a non-viral gene vector [40]. The CAP-functionalized PEI vectors showed specificity for cartilage tissue and gene transfection efficiency in the knee joints was demonstrated to be excellent, and can be employed for cartilage therapy. In another study, they employed the same construct to deliver siRNA into the cartilage of the knee joints to silence the expression of Hif-2α [41]. Hif-2α, which is one of the molecules that triggers cartilage degradation in osteoarthritis (OA), was therefore downregulated and cartilage degeneration and synovium inflammation in the knee joints were alleviated. In both cases, they showed that the use of the cartilage specific and chondrocyte-homing peptide identified by phage display could make therapy of degenerate cartilage feasible.

2.7 Using phage display for diagnosis and imaging

The development of osteoarthritis or rheumatoid arthritis is noted to be highly linked with MMP13 expression, a collagenase that degrades collagen and biglycans [42, 43]. Sun-Jun and his colleagues' utilized phage display to map out the substrate specificity of this enzyme, their screening revealed that MMP13 targeted with specificity to peptide substrates that have proline at the P3 position and lipophilic amino acids at P1′. They observed that a change in proline via site-directed mutagenesis made these substrates less sensitive to collagenase 3 [44]. Integrins are transmembrane heterodimeric proteins that play a role as mechanotransducers; they also mediate a number of other signaling cascades and triggers endocytosis [45] and or pinocytosis [46] that mediate cellular internalization. Chondrocytes plasma membranes have surface integrins subunits [47]. A fluorophore that is therefore bound to a ligand that interacts with integrins can be internalized. Hart and his coworkers demonstrated using bacteriophage Fd that displayed the cyclic integrin-binding peptide sequence GGCRGDMFGC on the major coat protein subunits. This led to the internalization of the phage by cells, thus demonstrating that the integrin-binding peptides displayed on the phage could target cells expression integrin on its surface for internalization [48], and this could be exploited to have the phage coat protein also functionalized to a fluorophore that serve for immunofluorescent imaging. Same process can also be exploited for the possible introduction of siRNA or preloaded drugs into cells for therapy.

2.8 Deploring CRISPR with phage technology for cartilage regeneration

CRISPR technology has proven to be a highly efficient and specific target genome editing technology for eukaryotes; and has been demonstrated as an excellent technology for specific genes silencing, genes knockouts, or knockdowns applications. Therefore this technology can be employed to silence specific gene products in cartilage tissues that amplify the deterioration of cartilage during rheumatoid arthritis or osteoarthritis. For instance, high MMP13 [49, 50], RUNX2 [51], VEGF [52] etc. expressions in cartilage tissue usually are pointers to cartilage degeneration and abnormality [53] and hence can be silenced through this CRISPR technology. The bottleneck remains nonetheless on homing CRISPR to cartilage tissue. Shefah and his team showed that P22 phage served as a robust supramolecular protein cage that could be utilized for cell type-specific delivery of encapsulated cargos [54]. They genetically fused Cas9 to a truncated form of the P22 phage scaffold protein, thereby packaging Cas9 and a single-guide RNA (sgRNA) inside the P22 capsid. Since the sgRNA is tunable, specifying which gene to target therefore is achievable. Homing such a delivery vehicle to cartilage tissue can be achieved via molecular engineering process. The chondrocyte-homing peptide, DWRVIIPPRPSA as discovered by Pi and his colleagues [40] could be chemically functionalized to the engineered P22 phage capsid construct (P22-Cas9: sgRNA complex) using polyethyleneimine. On the other hand, a genetic engineering approach in which the P22 phage tailspike protein is tinkered to contain this chondrocyte homing peptide (DWRVIIPPRPSA) especially at the C-terminus can then be assembled onto the P22 phage capsid construct encapsulating the Cas9-sgRNA. The P22 tailspike protein is well known for its tolerance to several physiological and environmental conditions such as protease, heat and detergents [55]. It is biocompatible and posse no harm to the human body. The non-covalent but irreversible binding of the phage's tailspike to its capsid will lead to the production of a phage construct with capsid loaded with the right gene regulatory factor(s) that has the capacity for cartilage specific targeting.

2.9 Using phage to develop biosensors for cartilage wound progression

Cartilage defect in knee such as in the case of rheumatoid arthritis or osteoarthritis or even in the event of joint injury can lead to matrix metalloproteinase expression enhancement and degenerative events [42, 43, 56, 57]. Inflammation is also known to be associated with joint symptoms and progression of osteoarthritis. The molecular markers of inflammation can be assessed in joint fluids and tissues from patients [58] using phage display technology. Phage based biosensors can be employed to sense the degeneration and extent of wound and even early detection of the degenerative event. These biosensors could reflect the effects of medical treatment. For instance, phage library can be screened against the MMP13 upregulation in osteoarthritis to select highly selective and affinity-binding phages to MMP13. Similar studies as done by Sun-Jun and his colleagues' as mentioned earlier utilized phage display to map out specificity to MMP13 [44]. These selective phages can detect MMP13 in the injured or degenerative joints and thus can be employed for sensor designs and constructs. Several phage-based biosensors have been constructed for detection of pathogens, antigens, secreted proteins in various disease states [59–61]. For instance, Singh and Amit used immobilized engineered tail spike proteins derived from the P22 bacteriophage onto gold surfaces using thiol-chemistry to analytical detect *Salmonella* with the sensitivity of 10^3 CFU/mL [62]. This technique has also been employed to successfully detect *E. coli* O157:H7, methicillin-resistant *S. aureus* [63], *S. aureus* [64], and hepatitis B virus [65]. Similarly, landscape phage has been successfully used as a molecular recognition interface to detect *Bacillus anthracic* spores [66], *Salmonella* [62, 67] and even in the detection of prostate serum antigen [68].

3. Conclusions

Even though there exist copious discoveries on the genetic factors as well as the molecular mechanisms surrounding cartilage degeneration, the efficacious treatment modalities remain elusive. Phage display provides an advantageous platform to study, diagnose and treat cartilage related diseases, since this provide a nano scale molecular mechanism that have the benefits of possessing higher tissue penetration, high specificities to cartilage, tunable, and hence can be leveraged for cartilage therapy, diagnoses, imaging and research application. By far, majority of phages used for display are biocompatible, and hence can serve as ideal drug delivery systems with minimal to no side effects to the human body upon administration, and should attain tremendous efficacy. The ability to fine tune drug loaded phages by functionalizing homing peptides to the phage particle offer a special pharmacokinetic characteristic, since it provides for regulated and targeted distribution of the payload, and ensure safety. Nonetheless, the use of phages for cartilage therapy still remains an obscured subject, and many obstacles should necessarily be surmounted. First, the choice of the right phage display libraries through phage biopanning is a critical step that will ensure the generation of the right ligand peptides for display. Secondly, there must be a concerted effort to direct cartilage studies to understand more cartilage targeting peptides and the specific genetic and molecular mechanisms that should be reversed in degenerated cartilage therapy process. The specific receptor target moieties, chondrocyte-ECM dynamic relationships, the biology of cartilage tissue ECM remodeling and ECM molecules secretion, deposition and recycling must all be understood *in vivo* to ensure enhanced application of phage technology for human cartilage regeneration.

Abbreviations

AKT	Ak strain transforming
CFU	Colony forming unit.
CRISPR	clustered regularly interspaced short palindromic repeats.
ECM	extracellular matrix.
ERK	extracellular-signal-regulated kinase.
FGF2	fibroblast growth factors 2 (basic).
HA	hyaluoronic acid.
HABP	hyaluoronic acid binding protein.
HAP	hydroxyapatite.
Hif-2α	Hypoxia Inducible Factor-2 alpha.
MMPs	matrix metalloproteinases.
MSCs	Mesenchymal stem cells.
PEI	polyethyleneimine.
RHAMM	receptor for hyaluronan mediated motility.
RHO	Ras homologous.
RUNX2	Runt-related transcription factor 2.

sgRNA	single-guide RNA.
siRNA	Small interfering RNA.
SMAD	small Mothers against decapentaplegic.
Src	sarcoma, a tyrosine kinase protein encoded by the *SRC* gene.
TGF-β1	Transforming growth factor beta 1.
VEGF	vascular endothelial growth factor.

Author details

Ayariga Joseph Atia[1*], Abugri Daniel Azumah[1], Bedi Deepa[2] and Derrick Dean[1]

1 Alabama State University, Montgomery, AL, USA

2 Tuskegee University, Tuskegee, AL, USA

*Address all correspondence to: ayarigajosephatia@yahoo.co.uk

References

[1] Ariga K, HillJP, LeeMV, VinuA, Charvet R, Acharya S. Challenges and breakthroughs in recent research on self-assembly. Science and Technology of Advanced Materials. 2008;9:1. DOI:10.1088/1468-6996/9/1/014109.

[2] Steinbacher S, Miller S, Baxa U, Budisa N, Weintraub A, Seckler R, Huber R. Phage P22 tailspike protein: crystal structure of the head-binding domain at 2.3 A, fully refined structure of the endorhamnosidase at 1.56 A resolution, and the molecular basis of O-antigen recognition and cleavage. J Mol Biol. 1997;11:865-880. DOI: 10.1006/jmbi.1997.0922.

[3] Crick FH, Orgel LE. The theory of inter-allelic complementation. J Mol Biol. 1964;8:161-165. DOI: 10.1016/s0022-2836(64)80156-x.

[4] Rakonjac J, Bennett NJ, Spagnuolo J, Gagic D, Russel M. Filamentous bacteriophage: biology, phage display and nanotechnology applications. Curr. Issues Mol. Biol. 2011;13:51-75. DOI: 10.1002/9780470015902.a0000777

[5] Mobasheri A, Rayman MP, Gualillo O, Sellam J, van der Kraan P, Fearon U. The role of metabolism in the pathogenesis of osteoarthritis. Nat Rev Rheumatol. 2017;13:302-311. DOI: 10.1038/nrrheum.2017.50.

[6] Guo Q, Wang Y, Xu D, Nossent J, Pavlos NJ, Xu J. Rheumatoid arthritis: pathological mechanisms and modern pharmacologic therapies. Bone research.2018;6: 15. DOI: https://doi.org/10.1038/s41413-018-0016-9

[7] van der Linden MPM, le Cessie S, Karim R, van der Woude D, Rachel K, Tom WJH, van der Helm-van MAHM. Long-term impact of delay in assessment of patients with early arthritis. Arthritis Rheum. 2010;62:3537-3546. DOI: 10.1002/art.27692.

[8] Borgia F, Giuffrida R, Guarneri F, CannavòSP.RelapsingPolychondritis: An Updated Review. Biomedicines, 2018;6:84. DOI: 10.3390/biomedicines6030084

[9] Foidart JM, Abe S, Martin GR, Zizic TM, Barnett EV, Lawley TJ, Katz SJ. Antibodies to type II collagen in relapsing polychondritis. N. Engl. J. Med. 1978;299:1203-1207. DOI: 10.1056/NEJM197811302992202.

[10] Pauli RM. Achondroplasia: a comprehensive clinical review. Orphanet J Rare.2019;14:1: DOI: 10.1186/s13023-018-0972-6

[11] Schumann JA, Parente JJ. Costochondritis. In: StatPearls [Internet]. Treasure Island (FL): StatPearls Publishing; 2020.Available from: https://www.ncbi.nlm.nih.gov/books/NBK532931/

[12] Proulx AM, Zryd TW. Costochondritis: diagnosis and treatment. Am Fam Physician. 2009;80:617-620. PMID: 19817327

[13] Amin RM, Andrade NS,Neuman BJ. Lumbar Disc Herniation. Current reviews in musculoskeletal medicine.2017;10:507-516. DOI: 10.1007/s12178-017-9441-4.

[14] Chow WA. Chondrosarcoma: biology, genetics, and epigenetics. F1000Research.2018;7:1826. DOI: 10.12688/f1000research.15953.1

[15] Khandeparkar SG, Joshi A, KhandeT,Kesari M. A rare case of giant soft tissue chondroma of the wrist: A cytopathological study with review of the literature. Journal of cytology.2014;31:40-43. DOI: 10.4103/0970-9371.130695.

[16] James CB, Uhl TL: A review of articular cartilage pathology and the use of glucosamine sulfate. J Athl Train. 2001;36:413-419. PMID: 16558667.

[17] Sophia FJ, Bedi A, Rodeo SA.The basic science of articular cartilage: structure, composition, and function. Sports Health. 2009, 1: 461-468. PMID: 23015907

[18] Zhang X, Blalock D, Wang J. Classifications and definitions of normal joints. Osteoarthritis-progress in Basic Research and Treatment. 2015.

[19] He T, Abbineni G, Cao B, Mao C.Nanofibrous bio-inorganic hybrid structures formed through self-assembly and oriented mineralization of genetically engineered phage nanofibers. Nano micro small. 2010; 6:2230-2235. DOI: 10.1002/smll.201001108.

[20] FratzlPHS, Gupta EP, Paschalis PR. Structure and mechanical quality of the collagen–mineral nano-composite in bone. Mater. Chem. 2004;14:2115.

[21] So YY, Jin-Woo O, Seung-Wuk L. Phage-Chips for Novel Optically Readable Tissue Engineering Assays. Langmuir. 2012; 28:2166-2172. DOI: https://doi.org/10.1021/la203840n.

[22] Zmolik JM, Mummert ME. Pep-1 as a novel probe for the in situ detection of hyaluronan. Journal of Histochemistry &Cytochemistry.2005;53:745-751. PubMed: 15928323.

[23] Campo GM, Micali A, Avenoso A, Ascola M, Scuruchi A, Pisani A, Bruschetta A, Calatroni D. Puzzolo S. Inhibition of small HA fragment activity and stimulation of A(2A) adenosine receptor pathway limit apoptosis and reduce cartilage damage in experimental arthritisHistochem. Cell Biol. 2015;143:531-543

[24] Binrui C, Yan L, Tao Y, Qing B, Mingying Y, Chuanbin M. Bacteriophage-based biomaterials for tissue regeneration. Advanced Drug Delivery Reviews.2019;145:73-95.

[25] Meng X, Jiangwei X, Gang W, Yu K, Liming F, Chunlin D, Hua L. Anchoring TGF-β1 on biomaterial surface via affinitive interactions; Effects on spatial structures and bioactivity. Colloids and Surfaces B: Biointerfaces. 2018;166:254-261.

[26] Hotchin NA, Hall A. The assembly of integrin adhesion complexes requires both extracellular matrix and intracellular rho/racGTPases. J Cell Biol. 1995;131:1857-1865.DOI: 10.1083/jcb.131.6.1857

[27] Cary LA, Han DC, Guan JL. Integrin-mediated signal transduction pathways. HistolHistopathol. 1999;14:1001-9. DOI:10.14670/HH-14.1001

[28] Clark EA, King WG, Brugge JS, Symons M, Hynes RO. Integrin-mediated signals regulated by members of the rho family of GTPases. J Cell Biol. 1998;142:573-586. DOI: 10.1083/jcb.142.2.573.

[29] Wang J, Gardner BM, Lu Q, Rodova M, Woodbury BG, Yost JG, Roby KF, Pinson DM, Tawfik O, Anderson HC. Transcription factor Nfat1 deficiency causes osteoarthritis through dysfunction of adult articular chondrocytes. J Pathol. 2009;219:163-172. DOI: 10.1002/path.2578.

[30] Alvarez R. SerraUnique and redundant roles of Smad3 in TGF-beta-mediated regulation of long bone development in organ cultureDevDyn. 2004;230:4: 685-699.

[31] HellingmanCA,Davidson EN, Koevoet W, Vitters EL, van den Berg WB, van Osch GJ. Smad signaling determines chondrogenic differentiation of bone-marrow-derived mesenchymal stem cells: inhibition of Smad1/5/8P

prevents terminal differentiation and calcification. Tissue Eng Part A. 2011; 17:1157-67. DOI: 10.1089/ten. TEA.2010.0043.

[32] Pollock JF, Healy K. Biomimetic and bio-responsive materials in regenerative medicine intelligent materials for healing living tissues. In: Santin M, editor. *Strategies in Regenerative Medicine.* New York, NY, USA: Springer. 2009; 97-154.

[33] Yue L, Alex AA, Robert L, Zhen G. Bioresponsive materials. Nat Rev Mater 2. 2017;16075. DOI: 10.1038/ natrevmats.2016.75

[34] Yoo SY, Kobayashi M, Lee PP, Lee SW. Early osteogenic differentiation of mouse preosteoblasts induced by collagen-derived DGEA-peptide on nanofibrous phage tissue matrices. Biomacromolecules. 2011;12:987-996.

[35] Yoo SY, Merzlyak A, Lee SW. Synthetic phage for tissue regeneration. Mediators Inflamm. 2014;192790. DOI: 10.1155/2014/192790.

[36] Ripellino JA, Klinger MM, Margolis RU, Margolis RK.The hyaluronic acid binding region as a specific probe for the localization of hyaluronic acid in tissue sections. Application to chick embryo and rat brain. J HistochemCytochem. 1985;33:1060-1066.

[37] Tolg C, Hamilton SR, Zalinska E, McCulloch L, Amin R, Akentieva N, Winnik F, Savani R, Bagli DJ, Luyt LG, Cowman MK, McCarthy JB, Turley EA, A RHAMM Mimetic Peptide Blocks Hyaluronan Signaling and Reduces Inflammation and Fibrogenesis in Excisional Skin Wounds. American Journal of Pathology. 2012;181:1250-1270. PubMed: 22889846

[38] Biao H, Qing L, Chao W, Pavan P, Sheila MA, BasakD, Hadi TN, RaminO, SiyuanZ, Christopher YL, Sherry LXX. Lucas L, MotomiE-I, Ling Q, Robert L,Mauck RV. IozzoDE, Birk LH. Decorin Regulates the Aggrecan Network Integrity and Biomechanical Functions of Cartilage Extracellular Matrix. ACS Nano. **2019;**13:11320-11333.

[39] Jarvinen TAH, Ruoslahti E. Target-seeking antifibrotic compound enhances wound healing and suppresses scar formation in mice. Proceedings of the National Academy of Sciences of the United States of America.2010;107: 21671-21676. PubMed: 21106754.

[40] Pi YB, Zhang X, Shi JJ, Zhu JX, Chen WQ, Zhang CG, Gao WW, Zhou CY, Ao YF, Targeted delivery of non-viral vectors to cartilage in vivo using a chondrocyte-homing peptide identified by phage display. Biomaterials. 2011; 32: 6324-6332. PubMed: 21624651.

[41] Pi Y, Zhang X, Shao Z, Zhao F, Hu X, Ao Y, Intra-articular delivery of anti-Hif-2 alpha siRNA by chondrocyte-homing nanoparticles to prevent cartilage degeneration in arthritic mice. Gene Therapy.2015;22:439-448. PubMed: 25876463

[42] Reboul P, Pelletier JP, Tardif G, Cloutier JM, Martel-Pelletier J. The new collagenase, collagenase-3, is expressed and synthesized by human chondrocytes but not by synoviocytes. A role in osteoarthritis. J Clin Invest. 1996; 97:2011-2019. DOI: 10.1172/ JCI118636.

[43] Billinghurst RC, Dahlberg L, Ionescu M, Reiner A, Bourne R, Rorabeck C, Mitchell P, Hambor J, Diekmann O, Tschesche H, Chen J, Van Wart H, Poole AR. Enhanced cleavage of type II collagen by collagenases in osteoarthritic articular cartilage. J Clin Invest. 1997;99:1534-1545. DOI: 10.1172/ JCI119316.

[44] Su-Jun D, Mark DB, Justin LM, Millard HL, Kevin RB, Luke HC, Jennifer N, Gregory P, Michael PW,

Marcia LM. Substrate Specificity of Human Collagenase 3 Assessed Using a Phage-displayed Peptide Library. JBC.2000; 275:40:31422-31427. DOI: 10.1074/jbc.M004538200

[45] Lee MY, Skoura A, Park EJ, Landskroner-Eiger S, Jozsef L, Luciano AK, Murata T, Pasula S, Dong Y, Bouaouina M, Calderwood DA, Ferguson SM, De CamilliP, Sessa WC. Dynamin 2 regulation of integrin endocytosis, but not VEGF signaling, is crucial for developmental angiogenesis. Development. 2014;141:1465-1472. DOI: 10.1242/dev.104539.

[46] Davis GE, Bayless KJ. An integrin and Rho GTPase-dependent pinocytic vacuole mechanism controls capillary lumen formation in collagen and fibrin matrices. Microcirculation. 2003;10:27-44. DOI: 10.1038/sj.mn.7800175.

[47] Dürr J, Goodman S, Potocnik A, von der Mark H, von der Mark K. Localization of beta 1-integrins in human cartilage and their role in chondrocyte adhesion to collagen and fibronectin. Exp Cell Res. 1993;207:235-244. DOI: 10.1006/excr.1993.1189.

[48] Hart SL, Knight AM, Harbottle RP, Mistry A, Hunger HD, Cutler DF, Williamson R, Coutelle C. Cell binding and internalization by filamentous phage displaying a cyclic Arg-Gly-Asp-containing peptide. J Biol Chem. 1994;269:12468-12474. PMID: 8175653.

[49] Meina W, Erik RS, HongtingJ, Jia L, Qiao HK, Hee-JeongIm, Chen D. MMP13 is a critical target gene during the progression of osteoarthritis. Arthritis Res Ther15, R5. 2013. DOI: 10.1186/ar4133

[50] Baragi VM, Becher G, Bendele AM, Biesinger R, Bluhm H, Boer J, Deng H, Dodd R, Essers M, Feuerstein T, Gallagher BM Jr, Gege C, Hochgürtel M, Hofmann M, Jaworski A, Jin L, Kiely A, Korniski B, Kroth H, Nix D, Nolte B, Piecha D, Powers TS, Richter F, Schneider M, Steeneck C, Sucholeiki I, Taveras A, Timmermann A, van Veldhuizen J, Weik J, Wu X, Xia B. A new class of potent matrix metalloproteinase 13 inhibitors for potential treatment of osteoarthritis: Evidence of histologic and clinical efficacy without musculoskeletal toxicity in rat models. Arthritis Rheum. 2009;60:2008-18. DOI: 10.1002/art.24629.

[51] Chena D, Dongyeon JK, JieS, Zhen Z, Regis JO. Runx2 plays a central role in Osteoarthritis development. Journal of Orthopaedic Translation. 2020;23:132-139.

[52] Janja Z, Peter V, Andrej C, Gregor H, Georges W, Sophie VS, Janja M. VEGF-A is associated with early degenerative changes in cartilage and subchondral bone. Growth Factors.2018;36:5-6, 263-273. DOI: 08977194.2019.1570926

[53] Lotz MK, Kraus VB. New developments in osteoarthritis. Posttraumatic osteoarthritis: pathogenesis and pharmacological treatment options. Arthritis Res Ther. 2010;12:211. DOI: 10.1186/ar3046.

[54] ShefahQ, Heini MM, Royce AW, Kimberly M, Trevor D, Blake W. Programmed Self-Assembly of an Active P22-Cas9 Nanocarrier System. Mol. Pharmaceutics. 2016;13:191-1196. DOI: 10.1021/acs.molpharmaceut.5b00822

[55] Joseph AA, Karthikeya V, Robert W, Hongzuan W, Doba J, Villafane R. Initiation of P22 Infection at the Phage Centennial, Frontiers in Science, Technology, Engineering and Mathematics. 2018;2: 64-81.

[56] Malemud CJ. Negative Regulators of JAK/STAT Signaling in Rheumatoid Arthritis and Osteoarthritis. Int J Mol Sci. 2017; 8:484. DOI: 10.3390/ijms18030484.

[57] Burrage PS, Mix KS, Brinckerhoff CE. Matrix metalloproteinases: role in arthritis. Front Biosci. 2006;11:529-543. DOI: 10.2741/1817.

[58] Lieberthal J, Sambamurthy N, Scanzello CR. Inflammation in joint injury and post-traumatic osteoarthritis. Osteoarthritis Cartilage. 2015;23:1825-1834. DOI: 10.1016/j.joca.2015.08.015.

[59] Vinay M, Franche N, Grégori G, Fantino JR, Pouillot F, Ansaldi M. Phage-Based Fluorescent Biosensor Prototypes to Specifically Detect Enteric Bacteria Such as E. coli and *Salmonella enterica* Typhimurium. PLoS One. 2015;10:7. DOI: 10.1371/journal.pone.0131466.

[60] Guo Y, Liang X, Zhou Y, Zhang Z, Wei H, Men D, Luo M, Zhang XE. Construction of bifunctional phage display for biological analysis and immunoassay. Anal Biochem. 2010;396:155-157. DOI: 10.1016/j.ab.2009.08.026.

[61] Zhang JL, Gou JJ, Zhang ZY, Jing YX, Zhang L, Guo R, Yan P, Cheng NL, Niu B, Xie J. Screening and evaluation of human single-chain fragment variable antibody against hepatitis B virus surface antigen. Hepatobiliary Pancreat Dis Int. 2006;5:237-241. PMID: 16698583.

[62] Singh, Amit. Immobilization of P22 Bacteriophage Tailspike Protein on Si Surface for Optimized Salmonella Capture. Analytical and Bioanalytical Techniques. 2013; S7. DOI: 10.4172/2155-9872.S7-007

[63] Nasser A, Azizian R, Tabasi M, Khezerloo JK, Heravi FS, Kalani MT, Sadeghifard N, Amini R, Pakzad I, Radmanesh A, Jalilian FA. Specification of Bacteriophage Isolated Against Clinical Methicillin-Resistant *Staphylococcus aureus*. Osong public health and research perspectives. 2019;10:20-24. DOI: 10.24171/j.phrp.2019.10.1.05

[64] Huang JX, Bishop-Hurley SL, Cooper MA. Development of anti-infectives using phage display: biological agents against bacteria, viruses, and parasites. Antimicrob Agents Chemother. 2012;56:4569-4582. DOI: 10.1128/AAC.00567-12.

[65] Tan WS, Ho KL. Phage display creates innovative applications to combat hepatitis B virus. World J Gastroenterol. 2014 Sep 7;20(33):11650-70. doi: 10.3748/wjg.v20.i33.11650. PMID: 25206271; PMCID: PMC4155357.

[66] Williams DD, Benedek O, Turnbough CL Jr. Species-specific peptide ligands for the detection of Bacillus anthracis spores. Appl Environ Microbiol. 2003;69:6288-6293. DOI: 10.1128/aem.69.10.6288-6293.2003.

[67] Wei S, Chelliah R, Rubab M, Oh DH, Uddin MJ, Ahn J. Bacteriophages as Potential Tools for Detection and Control of *Salmonella* spp. in Food Systems. Microorganisms. 2019;7:570. DOI: 10.3390/microorganisms7110570.

[68] Wu P, Leinonen J, Koivunen E, Lankinen H, Stenman UH. Identification of novel prostate-specific antigen-binding peptides modulating its enzyme activity. Eur J Biochem. 2000;267:6212-6220. DOI: 10.1046/j.1432-1327.2000.01696.x.

Section 3

Scope of Bacteriophage Therapy

Chapter 8

Therapeutic Efficacy of Bacteriophages

Ramasamy Palaniappan and Govindan Dayanithi

Abstract

Bacteriophages are bacterial cell-borne viruses that act as natural bacteria killers and they have been identified as therapeutic antibacterial agents. Bacteriophage therapy is a bacterial disease medication that is given to humans after a diagnosis of the disease to prevent and manage a number of bacterial infections. The ability of phage to invade and destroy their target bacterial host cells determines the efficacy of bacteriophage therapy. Bacteriophage therapy, which can be specific or nonspecific and can include a single phage or a cocktail of phages, is a safe treatment choice for antibiotic-resistant and recurrent bacterial infections after antibiotics have failed. A therapy is a cure for health problems, which is administered after the diagnosis of the diseases in the patient. Such non-antibiotic treatment approaches for drug-resistant bacteria are thought to be a promising new alternative to antibiotic therapy and vaccination. The occurrence, biology, morphology, infectivity, lysogenic and lytic behaviours, efficacy, and mechanisms of bacteriophages' therapeutic potentials for control and treatment of multidrug-resistant/sensitive bacterial infections are discussed. Isolation, long-term storage and recovery of lytic bacteriophages, bioassays, *in vivo* and *in vitro* experiments, and bacteriophage therapy validation are all identified. Holins, endolysins, ectolysins, and bacteriocins are bacteriophage antibacterial enzymes that are specific. Endolysins cause the target bacterium to lyse instantly, and hence their therapeutic potential has been explored in "Endolysin therapy." Endolysins have a high degree of biochemical variability, with certain lysins having a wider bactericidal function than antibiotics, while their bactericidal activities are far narrower. Bacteriophage recombinant lysins (chimeric streptococcal–staphylococcal constructs) have high specificity for a single bacterial species, killing only that species (lysin (CF-301) is focused to kill methicillin resistant *Staphylococcus aureus* (MRSA)), while other lysins have a broader lytic activity, killing several different bacterial species and hence the range of bactericidal activity. New advances in medicine, food safety, agriculture, and biotechnology demonstrate molecular engineering, such as the optimization of endolysins for particular applications. Small molecule antibiotics are replaced by lysins. The chapter discusses the occurrences of lytic phage in pathogenic bacteria in animals and humans, as well as the possible therapeutic effects of endolysins-bacteriophage therapy *in vivo* and *in vitro*, demonstrating the utility and efficacy of the therapy. Further developments in the bacteriophage assay, unique molecular-phage therapy, or a cocktail of phage for the control of a broad range of drug-resistant bacteria-host systems can promote non-antibiotic treatment methods as a viable alternative to conventional antibiotic therapy.

 IntechOpen

Keywords: bacteriophages, bacteria, endolysins, therapy, therapeutic effects, cocktails, antibiotic resistance, multidrug resistant bacteria, *in vivo*, *in vitro*, experiments, control

1. Introduction

Bacteriophages (phage) are bacterial viruses that are also known as 'natural killer phages' may take over their bacterial host and use it to grow and multiply. The phage may recognise, infect, and kill specific bacteria or groups of bacteria, as well as their host cells of unrelated bacteria. As a result, they play an important role in bacterial population regulation. Bacteriophages are used to (a) identify specific pathogens to help in pathogen detection and (b) destroy bacterial infections in a process known as lysogeny, in which one bacterium kills another through phage particles [1–4]. Since he first discovered bacteriophages in 1917, and later in 1919, a phage treatment was offered to cure a child suffering from dysentery, and the child was cured of the illness after a single dose of phage administration, D'Herelle is widely regarded as the father of bacteriophages. Since then, the phage cocktail's protection has been verified by administering it to a number of other healthy people [3, 4]. He also noted in 1919 that bacteriophages provided between chickens effectively reduce the mortality of chickens suffering from *Salmonella* infections, indicating that phage therapy experiments against bacterial infections were extremely successful [3–5]. D'Herelle published a comprehensive account of bacteriophages and founded "An International Bacteriophage Institute" in Tbilisi, Georgia, in 1923, which is now known as "the George Eliava Institute of Bacteriophages, Microbiology, and Virology" [3–5]. The Institute is engaged in the production and distribution of therapeutic bacteriophages for the treatment of a variety of bacterial infections. Bacteriophages have been successfully used to treat skin and diarrhoeal infections caused by *Staphylococcus aureus* and *Shigella dysenteriae* [6–8]. However, phage treatment has been poor since the discovery of antibiotics, large-scale development and availability, and widespread clinical use [9–13]. Furthermore, there was a chance of endotoxin contamination since most phage therapy trials lacked random and placebo controls [5]. Overuse and misuse of antibacterial drugs have been recorded since the dawn of the antibiotic era, resulting in intolerable antibiotic resistance with an approximate global intake of 100,000–200,000 tonnes of antibiotics per year [14, 15]. Antibiotic resistance in bacteria has arisen from such indiscriminate prophylactic use of multiple antibiotics, affecting all aspects of life and public health [2, 14–18]. Antimicrobial resistance is becoming a global threat, with the World Health Organization predicting that it could kill at least 50 million people every year by 2050 [19]. As antibiotic resistance rises, researchers are looking for new ways to detect and manage drug-resistant bacterial infections [1, 2, 4, 5]. Antimicrobial-resistant bacteria have evolved from bacteria with intrinsically drug-sensitive genes to bacteria with drug-resistant genes: Multidrug-resistant bacteria are classified as bacteria that are resistant to at least one antimicrobial agent out of three or more, while drug-resistant bacteria are defined as bacteria that are resistant to all antimicrobial stages. The advent and distribution of antimicrobials has increased rapidly due to widespread use of antibiotics as a supplement in animal husbandry, misuse of various antibiotics in clinics [2, 9–11, 13]. Antimicrobials' proliferation and dissemination have accelerated in tandem with international mobility. Existing antibacterial agents were unable to destroy bacteria immune to antibiotics, ushering in the "post-antibiotic" period [9, 14–18, 20–22]. Because of their specific antimicrobial activity as an alternative to antibiotics, bacteriophage treatment is gaining popularity as a means of ensuring future development. When antibiotics are ineffective against bacterial infections, phage therapy may help eradicate such complicated problems as a reliable treatment choice. In recent years, bacteriophages have been used to biocontrol bacterial numbers in agriculture, veterinary science, aquaculture, and the food industry [2, 10–13]. Bacteriophages have been used in agriculture to combat plant bacterial infections such as *Xanthomonas citri,* which would otherwise be treated with antibiotics. Holins, endolysin, ectolysin, and bacteriocins are bacteriophage antibacterial enzymes. Since endolysin targets induce immediate bacterial lysis, "endolysin therapy" has been developed to exploit their therapeutic potential [23]. Endolysin/recombinant endolysin has a lot of biochemical multiplication, and certain endolysins have a lot of bactericidal activity. Commercial applications have benefited from the use of endolysin enzymes or holins. The development of new drugs, creative methods, and the reduction of the risk of infectious agents and potential factors are all essential components of future bacterial disease control. Phage therapy reduces the development and replication of a wide variety of pathogenic bacteria, enhancing human and animal health and longevity. For particular groups of bacteria, however, the production of specific phage therapy cocktails is desirable. Phage therapy is a great way to treat microbial infections that are different depending on the operating system. Phage therapy is a fascinating rediscovered area of study that has many applications in science, agriculture, veterinary medicine, and medicine, including the potential prevention of antibiotic-resistant pathogens. The ability to combine antibiotic and phage therapy, the use of phage cocktails, and previously unexplored phage protein products are the most promising areas for the effective treatment of drug-resistant bacterial infections. Phage therapy is the subject of global research due to its wide range of applications and uses. This chapter addresses various aspects of phage therapy and how it can be used. After closely studying the protection and efficacy of phage, promising findings indicate that phage therapy against pathogenic bacteria could be the potential solution to pathogens that affect humans and animals.

1.1 Market potential of therapeutic bacteriophages

Bacteriophages are found all over the world, have many uses, and have contributed significantly to medicine, biotechnology, and molecular biology. Traditional antibiotic treatments are often replaced or supplemented by bacteriophage therapies and such alternative therapies has had a significant effect on revenues. In 2017, the global bacteriophage market was worth $567.9 million, and it is projected to grow at a 3.9% annual rate annual rate from 2018 to 2026. Globally, 600 million people are believed to be affected by foodborne diseases, with 420,000 people dying each year. In contrast, foodborne disease is said to affect 40% of children, resulting in 125,000 deaths per year [24]. The fastest-growing market will be for clinical applications of bacteriophages in phage therapy, diagnostics, drug development and manufacturing, phage display technology, antibacterial, vaccines, and biocontrol agents. Food and beverages currently hold the largest share of the global bacteriophage industry. Lytic bacteriophages are commonly used to control the spread of harmful infectious agents in foods such as fruits, vegetables, dairy products, and meals. Increased use of bacteriophages in such safe and healthy food items increased market potential. As a result, bacteriophages are being accepted for use in food safety applications in greater numbers. Companies are developing bacteriophage platforms and phagebanks (The Israeli Phage Bank (IPB) is a member of a global network of phage banks that provides a large assortment of purified bacteriophages) to treat multidrug-resistant bacteria in emergency situations [24]. Microgen, Amplify Bioscience Corporation, Ambiotics, and Phage Biotech Ltd. are some of the leading players in the bacteriophage industry. According to Amplify Biosciences Corporation, clinical trials for phage therapy against *Pseudomonas aeruginosa* infection in cystic fibrosis have begun in the United States [24].

1.2 Isolation and identification of bacteriophages

Seclusion, identification and propagation of the patient's infecting bacterial strain are critical for successful phage treatment. In medical practice, once a patient is suspected of having a contagious incurable infection, effective bacteriophages should be isolated, identified and purified from isolates of pathogenic bacteria occurring in the samples of urine, blood, and chronic wounds of patients. The bacterial colonies shall be picked up from selective Agar/LB agar plates according to their colony morphology, size, and pigmentation variability. The isolates are subjected to the staining procedures, biochemical and molecular tests and are cultured in various media to identify the genus and species of bacteria with the help of *Bergey's Manual of Determinative Bacteriology* and Bergey's manual of systematic bacteriology [25, 26]. Each of the bacterial isolates shall be transferred to LB broth at 37°C for 18 hours and then be stored at −20°C after the addition of 20% glycerol for further studies. With the development of diagnostic techniques, nonculture-methods such as 16S rRNA, PCR, RT-PCR, microarray, DNA/RNA sequencing, proteomics, ELISA and immunological methods and MALDI-TOP MS are used in clinical laboratories for microbial testing, identification and classification [1, 26–29]. If patients are opting for phage treatment, the foremost step is to isolate the disease-causing pathogenic bacteria using traditional methods. Subsequently the pathogens can be identified by using non-traditional methods. Bacteriophages uninfected host cells of bacteria multiply in Nutrient/LB agar plate to form a confluent film of bacterial growth over the surface of the plate at 37°C. In contrast, bacteriophages of infected cells of pathogenic bacterium if occur, bursts of such cells take place and release offspring bacteriophages. A visible, circular area of clearing zone in the confluent bacterial growth is known as a plaque, occurring after 8-10 hours of incubation and halos, zones of secondary lysis around plaques, can be identified after 24 hours. A suspension consisting of incubated samples of phage and cells of bacterial isolates shall be poured on to an appropriate LB/Nutrient agar medium to form a thin 'top layer'. Lastly, sensitivity and specificity of phage to the pathogenic bacteria is tested therapeutically. A key option in the treatment of infection is to use standard antibacterial drug therapy based on an anti-bacterial profile and / or physician experience [27]. Therefore, phage therapy will only be recommended if antibacterial drugs are unsuccessful and/or as soon as the infection is triggered by multidrug-resistant or pandrug-resistant bacteria.

2. Phage library and sensitivity

Availability of a library with a range of therapeutic bacteriophages is the foundation for the success of phage therapy. Bacteriophages are observed to show off a narrow to broad host specificity [30–32]. The lytic bacteriophages ought to have the capacity to kill strange bacterial species even as a range of bacteriophages can kill the identical bacterial strain (**Figures 1–4**) [33]. For a safe medical use of the phage, genes of toxins, antibiotic-resistance, and multidrug-resistant genes should not be present in the genome of the phage', whilst the lytic phage must have the potential to kill the multidrug-resistant bacteria such as *Acinetobacter baumannii, Enterococcus, Escherichia coli, Klebsiella pneumoniae, Pseudomonas aeruginosa,* and *Staphylococcus aureus* [34–38]. Confirmation of phage-sensitive bacteria is the prerequisite for

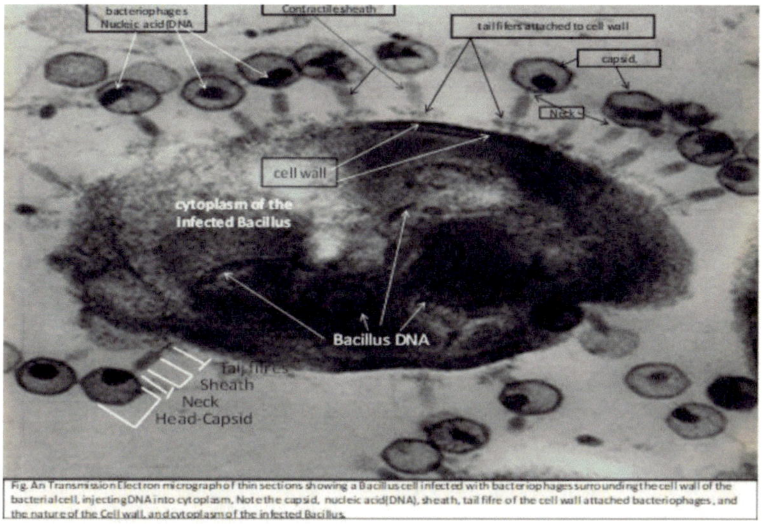

Figure 1.
Transmission electron micrograph (TEM) bacteriophages infected Bacillus. Note the capsid, nucleic acid, tail fibers [2].

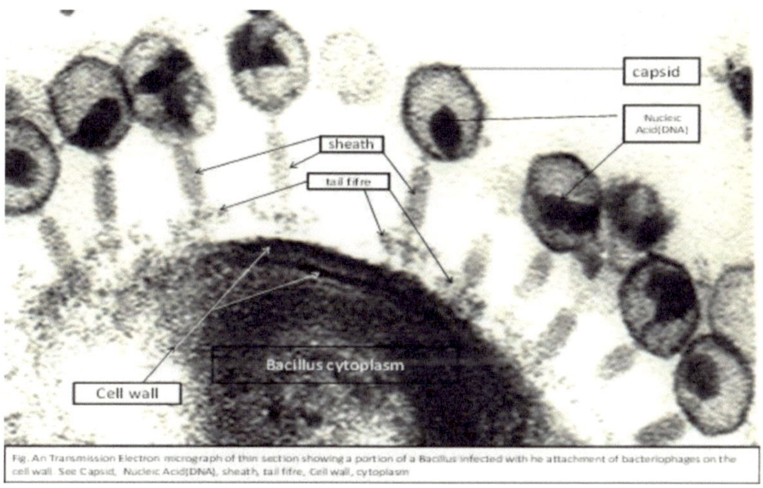

Figure 2.
A magnified portion of a Bacillus infected with bacteriophages is seen in this Transmission electron micrograph (TEM). The capsid, nucleic acid, tail fibers should all be noted [2].

initiation of antibacterial therapy. First, the bacterium inflicting the infection in the patient has to be received and identified; second, phage in the library that are effective against the pathogenic bacterium need to be screened, and selected for therapeutic use. If there are specific phage or cocktails of phage in the library that kill the identical bacterial strain, are preferred for the therapy [39, 40]. Reports have shown that phage cocktail preparations would possibly decorate bactericidal efficacy and additionally limit the chance of the emergence of phage-resistant isolates for the duration of the therapy [30–32]. Bacteriophages (phage) have a unique sorting mechanism for their target bacteria since they have a number of necessary characteristics such as inherent natural specificity, ease of use of cell signalling and receptor molecules, and simple phage or phage-derived product processing. These characteristics make bacteriophages more suitable for use as bacterial detectors and

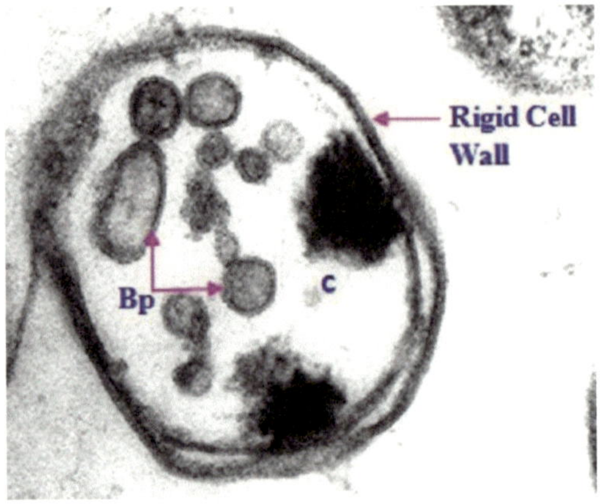

Figure 3.
TEM showing V. vulnificus *(VV-1) bacteriophage particles (Bp) within the cytolyzed cytoplasm (c) of the host cell bacterium* Vibrio *sp. Note the presence of phage within the cytoplasm [10].*

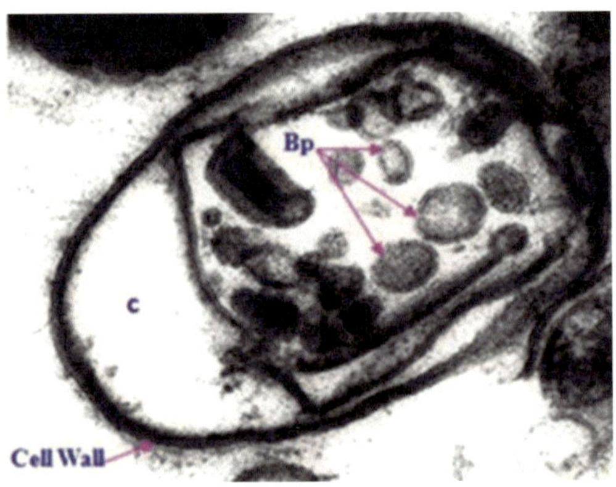

Figure 4.
Electron micrograph showing V. vulnificus *(VV-2) bacteriophage (Bp) particles within the lysed cytoplasm (c) of the host cell bacterium* Vibrio *sp [10].*

as aids in the detection of human pathogens. Phage-based systems are currently being used to diagnose *Staphylococcus aureus, Mycobacterium tuberculosis, Bacillus anthracis*, and *Yersinia pestis* in the clinical setting.

3. Production and purification of bacteriophages

Appropriate cultural media used for the growth, proliferation, and fermentation process of cells of bacterial hosts of bacteriophages for therapeutic applications. The fundamental processing of bacteriophages consists of several stages of purification (**Table 1**). These are broth specifications with low-speed centrifugation or filtering, cell removal, and cellular debris. Chloroform will be added to the lysate to form lysis and release the phage from non-lytic cells. Specified bacteriophage lysate can be used in many applications but clinical applications require additional purification of the lysate to eliminate endotoxins, metabolites, hydrophilic O-specific polysaccharide, phosphorylated oligosaccharides, phage, bacterial cells, and other wastes. Impure preparations of bacteriophages should not be used for injection [41, 47]. Occurrences of such as endotoxins in the bacteriophage preparations may aggravate

S. no	Therapeutic bacteriophages	Type of bacteria/ disease	Units of phages	Methods of isolation and purification	Therapeutic effects	Reference
1	Bacteriophages of *Pseudomonas, Klebsiella*, Serratia	Bacterial infection of *Pseudomonas, Klebsiella, Serratia*	A single production run can produce up to 64,000 treatment doses at 10^9 PFUs	Protocol used an aggregate of modified traditional techniques, membrane filtration processes, and no organic solvents to yield on average 23 mL of 10^{11} plaque-forming devices (PFUs)/ milliliter for *Pseudomonas, Klebsiella,* and *Serratia* phages tested	The protocol is beneficial for large scale standardized cultivation, purification, and manufacturing of bacteriophages. The approach emphases' eliminating endotoxins by up to 106-fold in phage preparations.	2020 Tiffany Luong [41]
2	T4-like coliphages or a commercial Russian coliphage	*E. coli* diarrhea Bacterial diarrhea	3.6×10^8 PFU of T4-like coliphage cocktail (T)	1 gm stool in 5 ml TS (8.5 g NaCl, 1 g tryptone/l), centrifuged at 14,500 g for 15 minutes. Filtered in a Millex AP20 prefilter and a Minisart filter. The presence of phages was determined in E. coli strains WG5 and K803, a K-12.	Fecal coliphage increased in treated over control children, however the titers did no longer show substantial intestinal phage replication. Lack of clinical efficacy of oral phages. No adverse events are attributable to oral phage application.	2016 Shafiqul Alam Sarker [42]
3	1. Bϕ-R2096 2.YMC 13/03/ R2096 ABA BP (phage Bϕ-R2096) lytic phages family Myoviridae	carbapenem-resistant *Acinetobacte baumannii* (CRAB) Serious nosocomial infection	Treated with concentrated phage Bϕ-R2096(1×10^{10} PFU) at two MOIs (MOI 100 and10)30 min after infection with CRAB (1×10^8CFU).	Carbapenem-resistant *A. baumannii* (CRAB) lytic phages isolated from sewage samples, purified, concentrated, treated NaCl (1M), (PEG) 8000 incubated at 4°C for 24 h, filtered using 0.22μm membranes and centrifuged at 12,000 × g for 1 h at 4°C, resuspended in sodium chloride-magnesium sulfate (SM) buffer.	Bacteria-only-infection group died rapidly. A sizable reduction in mortality in both *G. mellonella* larval and the mouse acute pneumonia models; Bϕ-R2096 improved the survival rates of each G. mellonella larvae and the mice *in vitro* and *in vivo*. No mortality or serious side effects in phage-treated experimental animal groups.	2019 Jongsoo Jeon, [43]
4	Lytic bacteriophage, phage 1513	Multidrug Resistance *Klebsiella pneumoniae* Pneumonia	Intranasal administration of a single dose of 2×10^9PFU/mouse 2h after KP 1513 inoculation	Centrifuged sewage sample, supernatant was supplemented with CaCl2. 10 mL supernatant, 10 mL 2 LB broth, and 2 mL bacteria solution [*K. pneumoniae* (KP 1513)] were incubated. By applying NaCl and (PEG8000), phages were precipitated,	Mice were protected from lethal pneumonia. When compared to untreated controls, phage-treated mice had a lower K. pneumoniae burden in the lungs. The phage KP 1513 has a significant antibacterial effect in vitro and in vivo, indicating that it could be	2014 Fang Cao [44]

S. no	Therapeutic bacteriophages	Type of bacteria/ disease	Units of phages	Methods of isolation and purification	Therapeutic effects	Reference
				ultracentrifuged, and passed via a Detoxi-Gel endotoxin removing gel. The purified phages were held at 4°C before they were used.	used instead of antibiotics to treat pneumonia caused by multidrug-resistant K. pneumoniae.	
4	A. baumannii phages	*Acinetobacter baumannii* Wound Infections	10^6 PFU. For cocktail synergy studies- 10^8 PFU per well for an MOI of 100	After growing AB5075 or AB5075P to exponential phase, 1 ml of each strain was added to 100-ml aliquots of the TSB-sewage mixture, inoculated with *A. baumannii,* incubated. 1 ml of infected TSB-sewage was centrifuged, the supernatant was filtered in a 0.22-m Spin-X centrifuge tube filter, and the centrifuged at 6,000 g. purified by using cesium chloride density centrifugation and filtered via a 0.22-m filter. Stocks of phages were held at 4° C.	The phages cocktail reduces bioburden in the wound, prevents infection and necrosis from spreading to adjacent tissue, and reduces infection-related morbidity.	2016 James M. Regeimbal [45]
5	CM8-1 and SJT-2 Bacteriophage	*Klebsiella pneumoniae* mastitis in dairy cattle	Bacteriophages bMECs were treated with or without K. pneumoniae (MOI, a 10:1 ratio of K. pneumoniae to bMECs), bacteriophages CM8-1, or SJT-2 (MOI, ratio of bacteriophage to K. pneumoniae was 1:10)	Bacteriophages CM8-1 and SJT-2 were isolated from dairy wastewater and mixed with a mid-log phase bacterial solution before being spread over a double-layer agar plate. Sodium magnesium (SM) buffer was applied after the bacteriophage had spread across the entire plate. A 0.22 m filter was used to filter the bacteriophage SM solution. PEG8000 (10%) was applied to the bacteriophage stock solution and stored at 4 o C overnight before being centrifuged for 10 minutes at 10,000 g.	Bacteriophages bMECs decreased bacterial adhesion, invasion, and cytotoxicity. The bacteriophage significantly reduced morphological damage and decreased TNF- and IL-1 concentrations, which were visible 4 to 8 hours after infection with *K. pneumoniae.*	2021 Yuxiang Shi, [46]

Table 1.
Methods of isolation and purification of therapeutic bacteriophages.

the immune system responses viz. fever, leucocytosis, leukopenia, fatal endotoxin shock, can open up macrophages, and release inflammatory mediators such as TNF-α, IL-6, and IL-1 and cause serious side effects. The final limit of endotoxins recommended for intravenous administration is 5 Endotoxin Unit (100 pg) (EU)/kg. The elimination of endotoxin from bacteriophages is a multidisciplinary procedure. Two-Phase fluid extraction processes viz. LPS affinity resins, ultrafiltration, and chromatographic methods for removal of well-charged endotoxin proteins. Ion exchange, size exclusion chromatography, interaction with histidine or polymyxin B, and anion-exchange chromatographic exchange were methods used for further phage purification. Diafiltration was used to exchange phage particles from lysate media with a suitable buffer. Cesium chloride density gradient centrifugation, ultracentrifugation, PEG precipitation, and ultrafiltration used for the removal of endotoxins and purification of phages. Bacteriophage CM8-1/SJT-2 stock solution mixed with bacterial culture in the mid-log phase spread on a double-layer agar Petri plate, Sodium magnesium (SM) buffer added after the bacteriophage had grown of the entire Petri plate and placed on a shaker at 120 rpm/min for 2 hours. The SM-bacteriophage lysate solution was filtered through a 0.22 μm filter, PEG8000 (10%) was once added to the bacteriophage stock solution, left the solution overnight at 4°C, and centrifuged at 10,000 × g for 10 min to obtain bacteriophage precipitation [46]. By contrast, T4 bacteriophages had prepared by using a stepwise gradient of anion-exchange quaternary amine (QA) CIM column and NaCl elution buffer [48, 49]. Purified bacteriophages of *Mycobacterium smegmatis* and *S. aureus* were prepared by using columns such as QA CIM and diethylamine (DEAE) while QA and DEAE CIM columns, were employed to remove endotoxins from pre-purified phage preparations by using the Endotrap HD column (Cambrex BioScience, EndoTraptm Blue) [50]. Enterococcal bacteriophages, viz. ENB6 and C33 were prepared from the raw wastewater by using caesium chloride density gradient centrifugation and stored at 4°C. Thus, there was a great deal of variation in the elution conditions between the different phages. A common operating procedure for varied phage preparations, storage, and transport is lacking [51]. Standard operating procedure (s) for large scale-bacteriophage cultivation, isolation, titration, and purification and to produce sufficient plaque-forming units of bacteriophages (PFUs) per milliliter of *Pseudomonas, Klebsiella, and Serratia* were established [41, 52]. Such a universal process and production of the final phage preparations for use could reduce endotoxins, might be pivotal in alleviating fears and the phage therapy shall be readily accepted all the world over.

4. Storage of phages

Phage preparations for clinical use ought to be (i) endotoxin-free, (ii) phage must be intact with high titers [53–55]. (iii) Suitable storage and transport are crucial. (iv) protected from high temperature, extremely acidic, or alkaline conditions [56], and (v) phage stock should not be refrozen and rethawed [57]. The usefulness of preparation of phage lysate, modified treatment methods evolved and accepted for long-term storage of phage was elucidated. In a study that demonstrated the infectivity of the phages remained unaffected with chloroform and DMSO treatments and storage for 30 days to a year at 4°C − 40°C [10, 58–60]. Infectivity of long-term stored bacteriophages at −80°C can be increased by adding 15%-25% glycerol to phage lysate preparations, and by rapid freezing and storage of phage infected bacteria at −70°C. Similarly, phage was shown to remain highly stable underneath normal storage conditions or also stable in NaCl and $MgSO_4$ due to its stabilizing effect. Considerable numbers of viable phage have been described to occur even after storage in distilled water. Phage isolates were found to remain stable upon storage at 4°C, or a rapid loss of phage infectivity was encountered with repeated freezing and thawing at −70°C. Phage infectivity could not be inhibited with trypsin, protease, ribonuclease treatments, or chloroform whilst the infectivity over the phage was inhibited together with lysozyme and SDS treatments [10, 59, 60]. The enzymatic treatments and inhibition of phage infectivity of several bacteriophages had been reported. Similarly, *Mycoplasma arthritidis* virulent 1 (MAV1) phage infectivity was reported to be unaffected by treatment with Triton X-100 and used to be resistant to non-ionic detergents [55, 61]. Phage survived a hundred percent at pH 7 and exhibited infectivity, whilst none of the phage survived at extreme pH conditions (pH 3 and pH 12) [10]. At a temperature below 37°C, phage JSF9 was shown to be stable whereas, at 50°C, the phage had been rapidly inactivated. Phage (VPP97) of *V. parahaemolyticus* have been shown to be stable up to 65°C and were totally inactivated at 70°C [10, 61–64]. Bacteriophages were detected to survive extremes over 95°C [52]. Bacteriophages such as T-ϕD0, T-ϕD2S, T-ϕHSIC, and T-ϕD1B exhibited a latent period ranging beyond 90°C [64]. The effects of temperature on the survival and infectivity of bacteriophages have clearly shown that the physicochemical parameters are very important for the survival and infectivity of phage [55, 58, 59]. Bacteriophages can be resilient to low/high temperatures, salinity, pH, and ions. They can tolerate extreme environments. New data on these along with therapeutic phage survivability, methods of their preservation and transport shall be useful.

5. Phage therapy

A bacteriophage therapy is a treatment for a patient's bacterial disease illness that is provided after the patient has been diagnosed. Bacteriophages are the most valuable and ubiquitous (10^{31}) organisms in the world, and are known to infect >140 bacterial genera. Description of phages and their antibacterial activity has initially been set up [6]. Bacteriophage therapy exhibits precise antibacterial lytic activities that have turned out to be a really useful concept to kill even an intracellular pathogenic bacterium and guarantee future development and consequently the therapeutic phages are re-emerging. As a substitute to antibiotics, experimental bacteriophage therapy might replace them when they fail to treat chronic infections, and such successful eradication of drug-resistant bacteria has been properly identified and demonstrated [65–72]. A single dose of phage has been shown to be more effective treatment than many doses of antibiotics such as amphetamines, tetracycline but chloramphenicol [73]. Moreover, careful phage collection, propagation, and purification requires complete experimental conditions. Such a focus ought to assist in the improvement of medical phage therapy utilized to a variety of systems, which is viewed an attribute on an emerging choice to antibiotic therapy and vaccination. The consequences of phage therapy are dependent on the plan of preparations and rout of administration of bacteriophage. The best possible administration route for phage preparations which should facilitate sufficient phages coming into direct contact with the bacteria. Routes of phage administration vary from oral, intravenous to multiple topical applications. There are different types of bacteriophage preparations developed to facilitate direct contact of the phage with the pathogenic bacterium for special bacterial infections and they are: (i) a phage powder, phage-containing lotion or dry gauze layer containing phages could be used for skin infections [74]. (ii) bacteriophages that have been sprayed dry become phages that can be inhaled as powder [75–77]. (iii) aerosolized phage preparations may be chosen for respiratory tract infections [76–78]. (iv) cream of phage preparations for skin infections. (v) injectable types of phage formulations [41, 47, 79]. (vi) phage infusion preparations may be considered for bloodstream infection [80, 81]. (vii) capsules containing phages (encapsulation/micro-encapsulation) that can protect particles from stomach acid inactivation should be preferred for gastrointestinal infections [75, 82]. An improved understanding of how synergic interactions of bacteriophages, cocktails with antibiotics impact bacterial infection is needed to stop unintentional inhibition of phage replication. Aerophages and IV phages each rescued 50% of animals from severe MRSA pneumonia. A mixture of aerophages and IV phages rescued 91% of animals, which was higher than either monotherapy or cocktail phage therapy [12]. Phage alone or a mixture of phages with antibiotics were treated against several bacterial infections in skin, blood, lung, and chronic otitis [36, 66, 80]. In contrast, other clinical reports have shown that some phages do not work due to constant infection and ETEC (Enterotoxic *Escherichia coli*) -complex diarrhoea [42, 83]. However, the prevalence of MDR bacteria is increasing, and our port drug portfolio is obsolete. The evolution of antibiotic resistance bacteria has thus become a major world health care problem. Clinical threats include MRSA, *Mycobacterium tuberculosis* and Vancomycin-Resistant *Enterococcus* (VRE) [84–86]. MDR bacterial infection is challenging and expensive to treat because of the increased resistance to all the antibiotics in practice. According to the Centres for Disease Control and Prevention (CDC), two million people are infected with antibiotic-resistant bacteria, and 23,000 people die each year in the USA from antibiotic-resistant bacterial infections. Prescribing antibiotics for the treatment of only standardized bacterial infections may slow down the process, but will not slow down the overall trend. Frequent use of antibiotics against diseases in humans and other organisms contaminates the environment and its cumulative effect on the development of antibiotic-resistant bacteria. As the number of antibiotic-resistant bacteria increases, alternative methods must be developed to effectively control them. Therefore, the use of antibiotics is a danger. Bacteriophage therapy with specific phages or a cocktail of phages signify an exciting alternative development to antibiotic therapy and vaccination. The progress of bacteriophage assays, biosensor tools, and bio-nano-targeted drug delivery system against drug-resistant bacteria elucidated. Bacteriophages are highly specific to target bacteria, and hence its usage is targeted toward a specific bacterial species and significantly minimizes off-targets effects on microbiome or human patient, as bacteriophages do not directly affect human cells [87]. Thus, phage treatment has been re-emphasized as the severity of drug-resistant bacteria has increased [66]. Therapeutic bacteriophages, units and outcome of the treatment of some antibiotic resistant bacterial infections are presented in **Tables 1** and **2**.

5.1 Personalized therapeutic phage

The term "personalised phage therapy" refers to the preparation and precise targeting of phage(s) against bacteria isolated from infected patients. Phage therapy has made extensive use of such a precise approach [80, 96], (**Table 3**). The patient's conditions need to be observed regularly to evaluate whether or not they are improving, and clinical samples from bacterial infection sites should be assessed in a timely manner to evaluate therapeutic efficacy, the emergence of phage-resistant strains and efficient phage titers. Phage should be replaced once a particular phage-resistant bacterium emerges [96]. If there are no phage in the library that kill a phage-resistant bacterial strain, the bacterial strain can be further used as a host bacterium to screen various types of samples (e.g., soils, faeces, urine) to isolate new effective phage. Such new bacteriophages can be added continuously to enrich the phage library if they meet the criteria. Phage therapy can be considered as an

S. no	Therapeutic bacteriophages	Type of bacteria/ disease	Units of phages used	Therapeutic effects	Reference
1.	Two novel bacteriophages, PBAB08 and PBAB25	(MDR) *Acinetobacter baumannii* Nasal infection	1×10^9 PFU of phage cocktail, intranasally injected	Mice treated with the phage cocktail showed a 2.3-fold higher survival rate than those untreated in 7 days post infection. 1/100 reduction of the number of *A. baumannii* in the lung of the mice treated with the phage cocktail.	2018 Kyoungeun Cha [88]
2.	PP1131 -phage cocktail	*Pseudomonas aeruginosa* Endocarditis	10^{10} PFU	Single-dose phage therapy was enough to control *P. aeruginosa* EE infections and act synergistically with ciprofloxacin. Phage-resistant mutants had impaired infectivity of *P. aeruginosa.*	2016 Frank Oechslin [89]
3.	Caudovirales phage strains, MPK1 and MPK6	Pseudomonas aeruginosa Peritonitis-sepsis caused by intraperitoneal (i.p.) infection	Mouse-2×10^6 or 2×10^7 PFU, *Drosophila melanogaster* - 5×10^7 PFU	Mice treated with phage had lower bacterial burdens in their livers, lungs, and spleens. Both phages significantly delayed the PAO1-induced killing of *D. melanogaster* ($P < 0.001$), although MPK1 persisted longer than MPK6 in uninfected D. melanogaster tissue samples. Infection is valid for evaluating the antibacterial efficacy of phage therapy against *P. aeruginosa* infections.	2009 Yun-Jeong Heo [90]
4.	Bacteriophage (MSa)	*Staphylococcus aureus*	Bacteriophage (MSa) (10^8 PFU)	All mice in the control group and the group treated with the lowest phage dose 107 PFU / mouse, died within 4 days (10/10 mice). mice treated with an intermediate dose 108 PFU/mouse were incompletely protected (2/5 mice survived). mice	2007 Rosanna Capparelli [91]

S. no	Therapeutic bacteriophages	Type of bacteria/ disease	Units of phages used	Therapeutic effects	Reference
				treated with the highest dose 10^9 PFU/mouse were all protected from the infection of *S. aureus*. The phage MSa inhibited abscess development.	
5.	A range of phages	*Pseudomonas aeruginosa* Chronic bilateral otitis externa	Approximately 400 PFU of phage (in 0.2 ml saline) were instilled into the right auditory canal.	No adverse effects were observed. *P. aeruginosa* was isolated from the ears after treatment, there were recurrent cycles of improvement and deterioration in the condition of the ears but they were better than before phage treatment	2006 J.A. Sivera Marza [92]
6.	Phage WSa	*Vibrio vulnificus*. Local and Systemic Disease	10^8 PFU	Infected mice with *V. vulnificus* may be treated to avoid local and systemic illness, as well as death. Phage therapy is a viable treatment choice for bacterial infections.	2002 Karen E. Cerveny [93]
7.	Enterococcus phages ENB6 and C33	Vancomycin-Resistant *Enterococcus faecium*. Gastrointestinal tract infection-VRE bacteremia and endocarditis	3×10^8 PFU of the phage strain	ENB6 phage formed plaques on 57% of the VRE clinical isolates and inhibited the bacterial growth of an additional 22% of the strains, thus exhibited an antibacterial effect against 79% of the strains. At higher doses of phage, 100% of the animals survived with minimal signs of illness such as mild lethargy in the first 24 hours	2001 Biswajit Biswas [94]
8.	Cocktail of four phages provided by Texas A&M and the San Diego-based biotech company AmpliPhi	Multidrug-Resistant Bacterial Infection. *Acinetobacter baumannii*	Phage cocktail is normally applied topically or taken orally. Phages were injected intravenously and into the abdominal cavity through catheters.	The bacteria gradually gained resistance to the phages, but the team compensated by constantly tweaking treatment with new phage strains and antibiotics, some of	2017 Scott LaFee and Heather Buschman [95]

S. no	Therapeutic bacteriophages	Type of bacteria/ disease	Units of phages used	Therapeutic effects	Reference
				which they had obtained from sewage. The road to recovery has not been without bumps. There have been setbacks that have nothing to do with the phages.	

Table 2.
Efficacy of therapeutic bacteriophages treatment of antibiotic resistant bacterial infections.

example of personalized medicine for bacterial infections [80]. Phage resistance may also be accompanied by changes in antibiotic resistance [99]. Therefore, the antibiotic resistance profile of phage-resistant strains should be simultaneously tested. The synergistic bactericidal activity of combining phage and antibiotics in the clinical cases should be considered [100] and further treatment strategies using phage alone and/or in combination with antibacterial drugs should be considered based on the results. The development of phage-sensitive and -resistant strains should be monitored regularly during phage therapy to see if phage therapy is a viable choice for successfully dealing with this issue. A clinical trial demonstrating the therapy's beneficial effects is critical in verifying its medical importance.

5.2 Gangrene wounds

Gangrene is the death of body tissue due to bacterial infection or lack of blood flow. Gas gangrene is caused by infection with a bacterium called *Clostridium perfringens* which in turn produces toxins that release gas causing tissue death [91]. A concoction of bacteriophages has been used to cure gangrene which is lively towards *Staphylococcus* spp., *Streptococcus* spp. and *Clostridium* [36, 50, 56]. Therapeutic efficacy of the phage has been improved, with the utility of "Pyophage" (a poly-specific cocktail of phage), achieved after detection of the particular etiologic agents and application of mono-specific lytic phage. The sequence of phage therapy treatments consisted of washings of the wound with a phage preparation, followed by subcutaneous injections of phage(s) as soon as to 4 instances per day. The utility of phage therapy has led to the removal of 69% Staphylococcal and 50% Streptococcal infections. Poly-specific (Pyophage, Sekstaphage) and mono-specific therapeutic phage cocktails developed have been used. Bacteriophages had been administered locally, via subcutaneous injections, and orally. Notably, phage therapy used to be carried out as a monotherapy, or complex treatment, which covered phage(s) and antibiotics administration. The investigations revealed that complicated treatment diminished the healing time by way of 1.2–2.5 times compared with antibiotic treatment. Even application of bacteriophages unique to one of the infectious agents in a wound expanded restoration and prompted quicker recuperation and purification. Importantly, it has been proved that a single utility of a bacteriophage would now not be adequate to stop infectious lesion problems. However, the investigators could not be concluding that the utility of bacteriophages barring antibiotics is better, as they had been unsuccessfully handled with antibiotics. They cautioned that the use of phage preparations supplied a fantastic impact on mono-infection, whilst complicated treatment, consisting of bacteriophages and

S. no	Therapeutic bacteriophages	Type of bacteria/ disease	Units of phages used	Therapeutic effects	Reference
1.	Φ2 (KpJH46Φ2)	*Klebsiella pneumoniae*. Prosthetic joint infection (PJI)	The patient received daily infusions of 6.3×10^{10} phages in 50 mL of normal saline each weekday for a total of 40doses.	Local symptoms, signs of infection, and recovery were all improved with phage therapy. The patient had no medication-related side effects and was asymptomatic 34 weeks after finishing treatment.	2020 Edison J Cano [81]
2.	Bacteriophage OMKO1	*Pseudomonas aeruginosa*. Prosthetic vascular graft infections	1,000 PFU phage OMKO1 (10e7 PFU/ ml) in 10 ml phage OMKO1	The infection tended to resolve after a single treatment of phage OMKO1 and ceftazidime, with no signs of recurrence.	2018 Chan, B. K. [97]
3.	Cocktail of 2 bacteriophages	Multidrug-resistant *Pseudomonas aeruginosa*, Bacteremia/sepsis after the ASD/ VSD closures	Dose of 3.5×10^{5} PFU every 6 hours.	When the patient resumed bacteriophage therapy, blood cultures that had reverted to positive for many days surprisingly, reproducibly reverted to sterile, which coincided with clinical progress.	2018 C. Duplessis [98]
4.	*A baumanii* bacteriophages	Multidrug-Resistant *Acinetobacter baumannii*. Craniectomy Site Infection	2×10^{10} PFU/mL, with an endotoxin level of 3.5×10^{5} endotoxin units (eu)/ mL. The phage dose given was 2.14×10^{7} PFU/mL	While the craniotomy site and skin flap healed well, fevers and leukocytosis continued. After surgical debridement, there were no more signs of infection at the craniotomy site, and no purulence to send for a repeat culture.	2018 Stephanie LaVergne [34]

Table 3.
Therapeutic bacteriophages for personalized treatments.

antibiotics, was once required for combined bacterial infections [101]. The use of distinctive bacteriophages was once greater than therapy with unique poly cocktails [102]. The most effective of this kind of custom-made phage therapy can be accelerated by using the specificity and virulence of phage to host strains. However, modified phage preparations require certain planning due to the fact they can incorporate temperate bacteriophages produced with the aid of a kind of scientific bacterium which has been used for adaptation.

5.3 Burn wounds

Burn wounds of patients have risks of bacterial infections. The floor of burn wound areas of sufferers may exhibit sepsis, lymphopenia, and intoxication. The

S. no	Therapeutic bacteriophages	Type of bacteria/ disease	Units of phages used	Therapeutic effects	Reference
1.	Lytic anti- *P aeruginosa* bacteriophages	*P. aeruginosa/ E. coli* Burn infections	PP1131; 1×10^6 PFU per mL	At very low concentrations, PP1131 decreased bacterial burden in burn wounds than standard of care	2017 Patrick Jault [83]
2.	*P. aeruginosa* phages 14/1 (Myoviridae) and PNM (Podoviridae) and *S. aureus* phage ISP (Myoviridae)	*Pseudomonas aeruginosa*, *Staphylococcus aureus*. Burn wound infection	10^9 PFU/ml of each phage	No adverse events, clinical abnormalities or changes in laboratory test results that could be related to the application of phages were observed.	2014 Thomas Rose [106]
3.	P. aeruginosa phages	*Pseudomonas aeruginosa* Burn Wound	10^8 PFU/100 µl inoculum of each of the following phages: Pa1 (ATCC 12175-B1); Pa2 (ATCC 14203-B1), and Pa11 (ATCC 14205-B1) (ATCC catalogue of bacteria and bacteriophages,	All the thermally injured mice that were not infected with PAO1Rif but administered the phage cocktail survived. The phage cocktail was not toxic to traumatized mice	2007 Catherine S. McVay [108]

Table 4.
Therapeutic bacteriophages for antibiotic-resistant burn infections.

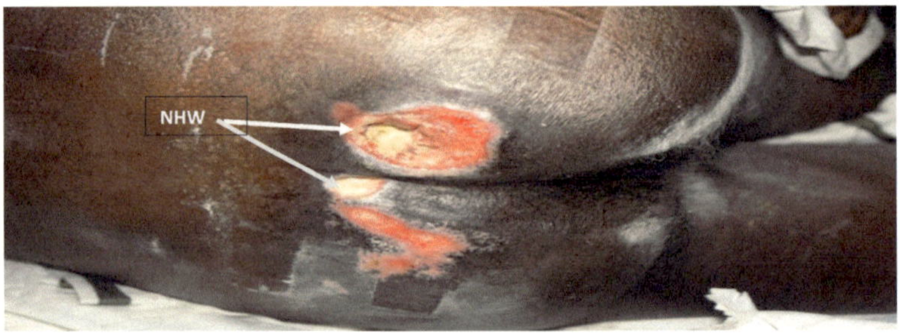

Figure 5.
Diabetic chronic non-healing wounds (NHW).

use of phage therapy was shown to be superb in eradication of pneumonia, the drug-resistant (MDR) *P. aeruginosa* infections in the burn wounds, and stopping the formation of sepsis [92, 103–107]. Therapeutic bacteriophages used for treatment of antibiotic-resistant burn infections are detailed in **Table 4**. In a complicated remedy comprising bacteriophages per OS and antibiotics, the use of bacteriophages has proven higher medical consequence in sufferers with contaminated burns (29% complicated instances of wounds) than in sufferers dealt with antibiotics (12.6% of cases) [103]. The volume of therapeutic phage particles ($\geq 10^6$ PFU/ml) used in the remedy is proven to be very extensive and the high-quality result of cure varied relying on the phage titer, routes of phage administration, sensitivity, specificity, and accessibility of bacterial host to the phage, length of phage therapy progression. A single dose (10^3 PFU/ml) of the phage BS24 has been confirmed to provide a wonderful impact and in contrast, no encouraging wound restoration response has been determined when the phage cocktail BFC-1 10^9 PFU/ml has been utilized at the wound floor [51, 92, 106, 109]. Dosage, remedy procedure, safety, efficacy, and pharmacodynamics of two phage cocktails, suggestions to deal with *E. coli*, and *P. aeruginosa* contaminated burn wounds are described [51].

5.4 Psoriasis

Psoriasis is a common chronic skin disease causing red and itchy scaly patches on the scalp, knees, elbows, and trunk. The "Phagoburn project" aims to reduce bacterial growth and reduce the incidence of psoriasis in patients with severe inflammation and infection. "Phagobon" has been used in the treatment of phage cures to treat *E. coli* and *Pseudomonas aeruginosa* diseases. Phase I/II clinical trials were established in France, Belgium, and Switzerland. Although this project is a breakthrough in phage medical studies, *in vitro* trials and clinics are still needed to gain widespread acceptance in the use of the therapeutic phage to treat people with pathogenic diseases or MDR. A biodegradable polymer wound dressing called, "PhagoBioDerm" is impregnated with numerous antimicrobial elements containing the phage cocktail Pyophage, and the dressing exhibited a slow degradation and presentation of the antimicrobial, and the release of phage particles for a long time had been demonstrated, exhibiting higher healing of infected venous leg ulcers [36–38, 96, 110]. The use of the PhagoBioDerm is promising for each remedy and prevention of microbial infections in wounds [36, 111]. Therapeutic bacteriophages, method of isolation, purification, and storage, units of phage, outcome of the bacteriophage therapy and reference are listed in **Table 1**.

5.5 Diabetes ulcers

Exposed non-healing wounds on the feet are considered "chronic ulcers". Chronic ulcers show up in sufferers with diabetes, atherosclerosis, and varicosity of the limbs (**Figure 5**). The healing processes of such chronic diabetic foot ulcers (DFU) depends on the coexisting infection of aerobic and anaerobic microorganisms' viz. *Staphylococcus* spp., *S. aureus*, *Proteobacteria*, and anaerobes *Anaerococcus*, *Bacteroides, Clostridium*, *Peptonihilus*, and *P. aeruginosa* [30, 31, 35–38, 110]. Antibacterial cure of ulcers infected with a variety of microbial organisms shall be difficult [32, 33, 35]. Long-term administration of antibiotics for healing the ulcers in diabetes mellitus sufferers may additionally be complicated and ineffective. In such complicated instances of infected diabetic foot ulcers, phage therapy could be an alternative or a supplementary treatment to antibiotics treatments. Phage therapy used to be the most incredible in ulcers with one bacterial agent (100%), however, a personalized phage therapeutic strategy can also lead to the removal of pathogens in instances with combined infections. There are quite a few studies that have described the efficacy and well-being of phage treatment of infected ulcers in humans. Previous antibiotic treatment was unsuccessful with a mixture of microbial infections of DFU unlike the results of the Phage therapy treatment of patients [38, 96]. The fundamental challenge in treating such infected wounds was once the inability to rapidly select phages towards all recognized bacterial diseases. Patients with DFU infected with methicillin-resistant and methicillin-sensitive *S. aureus* strains were effectively treated and cured with *Staphylococcus* phage Sb-1 [37, 111]. Commercially available phage cocktails can be chosen in every case following their specificity to particular infectious agents in an ulcer. When no such precise phage cocktail was once commercially available, a custom-made phage preparation can be prepared. Commercially available bacteriophage solution has been used against infections of *Escherichia coli, Pseudomonas aeruginosa, Staphylococcus aureus* occurring in Chronic Venous leg ulcers (VLU). No adverse events were reported for the study product and no significant differences were determined between the testing and control groups for the frequency of adverse events, the healing rate, or the frequency of healing [110].

5.6 Urinary tract infections

Acinetobacter baumannii is a Gram-negative nosocomial pathogen involved in human bacterial, meningitis, and respiratory infections (**Table 5**). A 68-year-old man with diabetes developed necrotizing pancreatitis, a complication of a pancreatic pseudocyst infected by the multi-drug resistant strain of *A. baumannii* [80]. Despite antibiotic treatment, the patient's condition deteriorated rapidly. Bacteriophage treatment has been initiated as part of an urgent new drug protocol. Commercially available Pyo bacteriophage solution (prophages; 20 mL) was used to enhance the treatment effect in the urinary tract infections in patients undergoing intravenous bacteriophage therapy TURP [112]. At very low concentrations of bacteriophage PP1131, the burden of the bacterium *P. aeruginosa* in burn wounds was less than the standard of care [83]. In another case of treatment, a solution consisting of 10^7–10^9 PFU/mL of the bacteriophages was introduced 2 times per 24 hours i.e., 8.00, 20.00 for 7 days, soon after surgery [113]. The patients were requested to hold the solution in the bladder for 30–60 min to control *Staphylococcus aureus, E. coli, Streptococcus* spp. (Streptococci group D renamed as *Enterococcus* spp.), *Pseudomonas aeruginosa*, *Proteus* spp. of urological infections of urinary tract infections after transurethral resection of the prostate. After treatment, four patients presented no significant bacterial growth while *E. coli* and *Enterococcus* spp. were still detected in the urine culture of four and one patient, individually. Bacterial counts decreased in six out of nine patients (67%), after the phage therapy treatment. No bacteriophage-associated adverse events have been detected. In one of the patients, (cephalosporin was given on day 3 after the development of fever (>38.0°C), the symptoms disappeared within 48 hours. Urine culture showed *P. aeruginosa* [113]. Intravascular bacteriophage therapy is no less than standard-protective care of antibiotic treatment, but it is no better than placebo bladder irrigation in terms of efficacy or safety in treating UTIs in patients with eruption. The data indicated that infection of the six lytic bacteriophages, each at a titre of 10 PFU mL − 1.20 mL ($\sim 2 \times 10^7$ p.f.u.) Pyo-phages were self-sustaining and self-limiting, with the phages decreasing in number along with the viable target organisms in which they replicated [114].

5.7 Pneumoniae

Bacterial pneumonia is an infection of *Streptococcus pneumoniae, Klebsiella pneumonia*, and *Mycoplasma pneumoniae* in each lung inflicting irritation in the alveoli or air sacs stuffed with fluid or pus, making it is hard to breathe. Pneumonia Phage (Φ2 (KpJH46Φ2)), *Klebsiella pneumoniae* examined for K Joint affected person against prostatic infection [43, 44], (**Tables 1** and **3**). The affected person received 6.3×10^{10} phages in 50 ml of normal saline solution and forty, doses every week. As a result of the treatment of phage, the local characteristics and *K. pneumoniae* infection symptoms had been resolved, the overall performance also had been restored. The affected person did now not experience any adverse effects related to treatment and remained asymptomatic within 34 weeks of completion of phage therapy when receiving minocycline. Intravenous injection of a single dose of 2×10^9 PFU of lytic bacteriophage of multidrug resistance *Klebsiella pneumoniae* KP

S. no	Therapeutic bacteriophages	Type of bacteria/ disease	Units of phages used	Therapeutic effects	Reference
1.	Pyo bacteriophage cocktail	Urinary tract infections	Intravesical Pyo bacteriophage (Pyophage; 20 mL) or/intravesical placebo solution (20 mL) twice daily for 7 days in a double-blind fashion	Intravesical bacteriophage therapy was found to be comparable to regular antibiotic therapy. In terms of eptitude and protection in treating UTIs in patients undergoing TURP, however, it was not superior to placebo bladder irrigation.	2021 Lorenz Leitner et al., [112]
2.	Pyo bacteriophage	*Staphylococcus aureus, E. coli, Streptococcus* spp. (Streptococci group D renamed as *Enterococcus* spp.), *Pseudomonas aeruginosa*, *Proteus* spp. urinary tract infections after transurethral resection of the prostate	The bacteriophages have 10^7–10^9 PFU/mL. The solution was given twice every 24 hours, at 8 a.m. and 20 a.m., for seven days, beginning the day after surgery. The patients were advised to keep the solution for 30–60 minutes in their bladders.	Four patients had no noticeable bacterial growth after treatment, while *E. coli* and *Enterococcus* spp. were still contained in the urinary cultures of four and one patient, respectively. After phage therapy, bacterial counts decreased in six out of nine patients (67 percent). There have been no reported side effects. The symptoms in one of the patients vanished within 48 hours after receiving cephalosporin on day three after developing a fever (>38.0°C). *P. aeruginosa* was discovered in urine culture.	2018 Aleksandre Ujmajuridze [113]
3.	*A. baumannii*-specific lytic bacteriophage cocktails PC, IV, and IVB, PC AC4, C1P12, C2P21, C2P24	*A. baumannii*. Necrotizing pancreatitis complicated by an MDR *A. baumannii* infection	Average endotoxin levels of bacteriophage cocktails PC, IV, and IVB were $2.4x10^3$ EU/ml, $5.89x10^3$ EU/ml, and $1.64x10^3$ EU/ml, respectively 4 $x10^9$ PFU of bacteriophages, $5x10^9$ PFU of bacteriophages PC AC4, C1P12, C2P21,	The administration of bacteriophages intravenously and percutaneously into the abscess cavities was linked to the patient's clinical course being reversed, clearance of the *A. baumannii* infection, and a returning to health.	2017 Robert T. Schooley [80]

S. no	Therapeutic bacteriophages	Type of bacteria/ disease	Units of phages used	Therapeutic effects	Reference
			C2P24, Intracavitary, Intravenous 2		
4.	Lytic bacteriophages	*Pseudomonas aeruginosa*, Urinary tract infection	Six lytic bacteriophages, each at a titre of 10^6 PFU ml − 1 … .20 ml (~2×10^7 p.f.u.) Pyophage	The bacteriophage infection was self-sustaining and self-limiting, with the phage's number declining in tandem with the number of viable target species in which it replicated.	2011A. Khawaldeh [114]

Table 5.
Therapeutic bacteriophages for antibiotic-resistant Urinary tract bacterial infections.

1513/mouse protected animals from sublethal pneumonia. The severity of pneumonia has been shown to be low. Compared with the untreated control, Phage-treated mice are more unlikely to develop *Klebsiella pneumoniae* in the lungs. Phage KP 1513, has a significant antibacterial impact *in vitro* and *in vivo*, and its host *K. pneumoniae* is multi-drug-resistant. Phage KP 1513 can be used as choice to antibiotic treatment for pneumonia caused by *Klebsiella pneumoniae* [44]. Aerophages/ Intravenous injection of bacteriophages saved 50% of animals from severe MRSA pneumonia compared to placebo controls. In contrast, administration of bacteriophages by both the aerophages and IV phages rescued 91% of animals, which used to be greater than either monotherapy. Standard-of-care antibiotic linezolid saved 38% of animals [79]. The natural phages belonging to Caudovirales including order Siphoviridae, Myoviridae, and Podoviridae had been separated from the clinical strains of multidrug-resistant *K. pneumoniae*. *In vitro* lytic activity of phages on isolated bacteria revealed 70% coverage of 33 isolated antibiotic-resistant strains, of which 50% targeted multiple phages. Overall, these results suggest the possibility of phage detection by strong action against antibiotic-resistant KP strains and may furnish a new therapeutic approach to the treatment of ESBL and CRKP infections [115].

5.8 Diarrhoea

Bacterial diarrhea occurs in humans if infected with bacteria such as *Salmonella* and *E. coli*. Symptoms of diarrhea appears if the lining of the intestine is unable to absorb fluid, or secretes fluid, and bowel activities become loose or watery 3 or more times a day. Loss of fluid and electrolytes were encountered as a result of diarrhea [42]. In a placebo-controlled clinical trial, oral administration of Coliphage 10^9 PFU against *Escherichia coli* 3 times/day/4 days showed no significant clinical benefit between the control and test group (**Table 1**) [51, 83]. Fifteen healthy volunteers with *Escherichia coli* diarrhea received *Escherichia coli* phage T4 dose (10^3 PFU/ml), high-phage dose (10^5 PFU/ml), and fifteen healthy adult volunteers received low dose *Escherichia coli* phage (PG4). Volunteers receiving high-dose (10^5 PFU/ml), high-dose phage showed stool phage 1 day after exposure. This prevalence is only 50% in those receiving low-dose bacteriophages. One week after the 2-day course of oral phage application, no faecal phage was detected. Oral phage application did not reduce the total stool *E. coli* count. In addition, no significant phage T4 replication was found in the early *E. coli* population. The study described the production of phage cocktails for use in clinical trials and Phage preparations are already entering clinical trials [51, 83]. Phage therapy has recently been re-emphasized due to the severity of drug-resistant bacterial infections [9]. Antibiotics alone or with antibiotics have been used successfully to treat a variety of bacterial infections, including atherosclerosis, lung and lung infections, chronic otitis, skin burn infections and enteric infections [65, 68, 80, 92, 116]. In contrast, other clinical reports have shown that bacteriophages are less effective than expected due to inadequacy or coverage for topical bacterial infections and ETEC (Enterotoxigenic *Escherichia coli*) [42, 83]. In addition, published reports show no side effects in clinical trials or no adversative actions associated to phage application [51, 117].

5.9 Tuberculosis

Tuberculosis (TB) is a lung infection caused by the endogenous bacterium *Mycobacterium tuberculosis*. First-line TB drugs drugs like rifampicin and isoniazid are resistant to certain types of multidrug-resistant tuberculosis (MDR-TB). In 2018, 484,000 new TB patients failed to respond to rifampicin, according to the World Health Organization (WHO). Seventy-eight percent of these patients have tuberculosis with multidrug resistance (MDR-TB) [118]. Mycobacteria exist in over 170 distinct species, each with its own pathogen development in humans [119]. *Mycobacterium ulcerans* and *M. leprae*, in addition to tuberculosis, *Mycobacterium* ulceration and *Mycobacterium* leprosy, respectively, cause Barley ulcers and leprosy [120]. Alternative therapies for MDR-TB are important for disease control, particularly as newer approaches to mycobacteriophage therapy emerge. To date, 11,282 mycobacteriophages have been discovered [121]. *M. smegmatis*, a non-pathogenic vector, can transport phages to the same intracellular compartments as *M. tuberculosis* [122]. M. mycobacteriophage D29's antimicrobial value was doubled after it was given twice in a 24-hour period to treat tuberculosis H37RV [123]. Aerosolized bacteriophage D29 treatment reduced TB cases in the lungs and vaccinated mice against tuberculosis [124]. Tuberculosis-prone health workers can benefit from aerosolized mycobacteriophages. D29 was used to treat burly ulcers caused by Mycobacterium ulcers in the Marine Footpad model [125]. As the disease progresses, infected patients experience necrosis of the skin, subcutaneous tissue, and bone, necessitating surgical skin rupture. Mycobacterial and pathological counts were decreased after D29 was injected subcutaneously. In footballs and lymph nodes, it causes the development of water-borne cytokines. This approach was used to deliver the lytic mycobacteriophage TM4 to *M. tuberculosis*-infected RAW264.7 macrophages, which decreased bacterial counts. On the other hand, the phage was found to be inactive on its own. *M. smegmatis*-TM4 complex substantially reduced bacterial counts in *M. avium*-infected mice's spleens, while TM4 or *M. smegmatis* alone had no effect [126]. Phage cocktails may be used to overcome phage resistance tuberculosis.

5.9.1 Endolysin therapy

Endolysin therapy is a major part of phage therapy. Endolysins are considered protein-based antibiotics or antimicrobials. The purified endolysin is a powerful antibacterial agent for curing bacterial infections in human beings and animals. The efficacy of endolysin enzyme, host bacteria, bacterial disease and the therapuetic use in experimental animal models is listed in **Table 6**. Endolysin is an enzyme used by the bacteriophages to degrade the bacterial host's peptidoglycan from the inside,

S. no	Enzyme	Sources/bacteriophages	Bacteria/disease	Unit of enzyme	Therapuetic effects	Reference
1.	Depolymerase KP34p57	Lytic phage KP34	*K. pneumoniae*	KP34p57 depolymerase was evaluated at four different concentrations: 7.5 ng/ml, 75 ng/ml, 750 ng/ml, and 7500 ng/ml.	After 2 hours of incubation with the enzyme, none of the concentrations examined showed major improvements in colony count.	2020 Latka, A., [127]
2.	Two capsule depolymerases (Dpo42 and Dpo43)	Phage IME205 isolated from a raw sewage	Carbapenem resistant *Klebsiella pneumoniae* (CRKP)	Dpo42 or Dpo43 (20 ng)	Both Dpo42 and Dpo43 depolymerases rendered the host bacteria prone to serum complement killing. Anti-virulent capsule depolymerases show promise in the battle against CRKP infections.	2020 Liu, Y., [128]
3.	LysSS	LysSS recombinant protein was purified from the host cells of *E. coli* BL21 Star® (DE3)	*Salmonella* spp., *A. baumannii*, *E. coli*, *P. aeruginosa*, *E. faecium*, *S. aureus*, *K. pneumoniae*	MHB was used to incorporate freshly grown bacteria (104 CFU/well) and purify them. LysSS was added at different concentrations (20–200 g/well) along with 200 L of 1 MHB per well in each well. 1.25 mg/mL LysSS was applied at the end.	Without pre-treatment with an outer membrane permeabiliser, LysSS demonstrated activity against MDR *A. baumannii*, MDR *E. coli*, MDR *K. pneumoniae*, MDR *P. aeruginosa*, and *Salmonella* sp. LysSS stopped methicillin-resistant *S. aureus* from growing (MRSA).	2020 Kim, S., [129]
4.	Tripleacting staphylolytic peptidoglycan hydrolases (Lysostaphin and LysK-)	Recombinant phage lysin proteins	*Staphylococcus aureus*	Lysostaphin 0.77 g/ml (27 nM); LysK, 47 g/ml (840 nM); Lysostaphin and LysK (L + K) in combination 0.2 g/ml (7 nM and 3 nM, respectively); triple fusion K-L, 7 g/ml (97 nM); triple fusion L-K, 7.8 g/ml (107 nM); triple fusion L-	Nasal colonisation was decreased by 87 percent when 200 g lysostaphin was used. In cultured mammary epithelial cells and a mouse model of *S. mastitis*, *S. aureus* demonstrated biofilm eradication and the ability to destroy intracellular *S. aureus*	2016 Becker, S. C., [130]
5.	PlyC holoenzyme, mediated by PlyCB subunit	C1 bacteriophage	*Streptococcus pyogenes* strain D471	Treatment with 50 μg/ml WT PlyC	Lower concentrations showed a dose response and reduced intracellular colonisation (CFUs) by 95% within 1 hour. The endolysins	2016 Shen, Y., [131]

S. no	Enzyme	Sources/bacteriophages	Bacteria/disease	Unit of enzyme	Therapeutic effects	Reference
					from the B30 and Ply700 streptococcal phages, on the other hand, failed to reduce intracellular Spy CFUs significantly.	
6.	Chimeric lysin	Phages infecting Gram-positive bacteria	methicillin-resistant *Staphylococcus aureus*. Burn wound infected with MRSA WHS11081	ClyF doses are 25, 37.5, and 50 mg/kg, with a maximum dose of 100 mg/kg. 6×10^7 CFU WHS11081 6×10^7 CFU WHS11081 /mouse model of MRSA WHS11081 infected burn wound (1×10^7 CFU/mouse)	Chimeric lysin is an effective antibacterial against MDR *S. aureus* in a mouse burn wound.	2017 Yang, H [132]
	λSA2-E-Lyso-SH3b and λSA2-E-LysK-SH3b	Bacteriophage endolysins (peptidoglycan hydrolases)	*Staphylococcus aureus*	100 μg/ml, λSA2-E-Lyso-SH3b and λSA2-E-LysK-, infusion of 25 μg of λSA2-E-Lyso-SH3b or λSA2-E-LysK-SH3b	Compared to control, SA2-E-Lyso-SH3b and SA2-E-LysK-SH3b decreased *S. aureus* bacterial load by 3 and 1 log units within 3 h at 100 g/ml, respectively. When SA2-E-LysK-SH3b and lysostaphin (12.5 g each/gland) were tested together in mice, they resulted in a 3.36-log reduction in CFU.	2012 Schmelcher, M. [23]
7.	Chimeric Lysin (ClyS)	Staphylococcus-specific phage	Methicillin-Resistant and -Sensitive *Staphylococcus aureus* strains	Topical ClyS activity was tested in vitro by combining 1 ml of PBS with a dose of 10% ClyS in Aquaphor. The mixture was centrifuged for 10 minutes at 4,000 rpm.	ClyS killed more methicillin-susceptible bacteria (MSSA) and methicillin-resistant S. aureus (MRSA) than mupirocinl, with a 2-log reduction with mupirocinl compared to a 3-log reduction with ClyS. In vitro, the use of ClyS reduced the ability for MRSA and MSSA species to develop resistance relative to the use of mupirocin.	2011 Pastagia, M. [133]

S. no	Enzyme	Sources/bacteriophages	Bacteria/disease	Unit of enzyme	Therapuetic effects	Reference
8.	ClyS	Fusion of N-terminal catalytic domain of *S. aureus* Twort phage lysin with C-terminal cell wall-targeting domain from another *S. aureus* phage lysin (phiNM3)	Methicillin-Resistant *Staphylococcus aureus*	A single treatment with ClyS, 200 U/mg or 7.1 U/nM One intraperitoneal dose of ClyS, ClyS and oxacillin at doses (A unit of ClyS activity per millilitre was described as the reciprocal of the highest lysin dilution that reduced absorbance by 50% in 15 minutes.	In a mouse nasal decolonization model, a single treatment with ClyS decreased the viability of MRSA cells by two logs in one hour. MRSA-infected septicemia mice were given a single intraperitoneal dose of ClyS and survived. In a mouse model of MRSA septic death, ClyS, in addition to oxacillin, offered synergistic defence against septic death.	2010 Daniel, A [134]
9.	Endolysin LysH5 and nisin	Staphylococcal bacteriophage phi-SauS-IPLA88	*Staphylococcus aureus*	300 nM Lys109	When LysH5 was combined with nisin, a bacteriocin, a significant synergistic effect was observed. Nisin and LysH5 minimum inhibitory concentrations were decreased by 64 and 16 times, respectively. On cell suspensions, Nisin increased LysH5's lytic activity by an order of magnitude.	2010 García, P [135]
10.	Pneumococcal lysins	Streptococcal bacteriophage	Antibiotic-resistant S. pneumoniae/ acute otitis media (AOM), septicemia,bronchitis, meningitis	A 2,000 µg2,000-µg dose of Cpl-1, mouse	Cpl-1-treated mice survived a fatal pneumonia infection 100 percent of the time. At 24 hours after infection, treated mice recovered quickly.	2009 Witzenrath, M. [117]
11.	Lysin Ply700	*Streptococcus uberis* (ATCC 700407)	*Streptococcus uberis*, *Streptococcus pyogenes*, *Streptococcus dysgalactiae*	Ply700 (50 µg/ml)	Activity against *E. coli*, *S. aureus*, or *S. agalactiae* induced a rapid, calcium-dependent lysis. *S. uberis* is killed. With an inoculating dose of 4500 cfu/ml, 31 percent killing was observed, while 81 percent killing	2008 Celia, L. K. [136]

S. no	Enzyme	Sources/bacteriophages	Bacteria/disease	Unit of enzyme	Therapuetic effects	Reference
					was observed when the inoculum was reduced to 600 cfu/ml.	
12.	PlyGBS	Streptococcal bacteriophage	Prophylactic for Group B streptococcal (GBS) vaginal colonization in pregnant women	Single dose of PlyGBS, 100 μl of purified PlyC	Single dose of PlyGBS could cause a 3 log10 reduction of the bacterial cellsnin mice that had been vaginally challenged with GBS. PlyGBS can be used as a decontaminant to eliminate GBS from new-borns. Can reduce the rate of neonatal meningitis and sepsis	2005 Cheng Q. [137]
13.	PlyV12	*E. faecalis* phage	VRE (Vancomycin-Resistant Enterococcus)	100 μl of PlyV12 at 25 U/ml.	The lysin exhibited lytic activityagainst *Staphylococcus* and Groups A, B and C streptococci, thus the lysin exhibited a broad lytic spectrum.	2004 Yoong, P. [138]
14.	PlyC lysin	Streptococcal bacteriophage C1	Groups A, C and E streptococci. Group A streptococci (GAS) S. pyogenes/ pharyngitis rheumatic fever	10 ng of PlyC, mouse	PlyC lysin successfully eliminated upper respiratory colonization of GAS in mice. None of the lysin treated mice were colonized compared to 100% of the control mice.	2001 Nelson, D. [139]

Table 6.
Endolysin therapy: therapeutic effects of Enzymes on bacterial diseases.

resulting in the release of cell lysis and offspring virions [23, 115]. *In vitro* and in mice models, the recombinant phage-derived lysins exhibit highly efficient bactericidal activity against multidrug-resistant *E. faecalis*. Endolysin LysEF-P10, EF24C, Lys168, Lys170 PlyV12 LysEF-P10, IME-EF1, and lysine CF-301 zap methicillin-resistant *Staphylococcus aureus* (MRSA). Antibiotic-resistant *S. pneumoniae*/Acute otitis media (AOM) infected mice exhibited a quick recovery from infection after 24 hours when they were treated with CPL-11 (therapeutic pneumococcal lysin streptococcal bacteriophage) at a dose of 2,000 μg [117]. The mixture of lysostaphin and the chimeric phage lysin λSA2E-LysK-SH3b synergistically kill *S. aureus in vitro* and in mouse models of bovine mastitis [140, 141]. Some lysins, such as CF-301, N-Refasin, P128, and Art-55, are at various stages of pre-clinical or clinical development and are antibacterial for the cure of multiple antibiotic drug-resistant (MDR) infections of Gram-positive, and Gram-negative pathogens. Synchronization of pneumococcal phage lysine with CPL-1 and autolysin LytA eliminates *Streptococcus pneumonia, S. pseudopunemonia,* and *S. aureus* [80, 140]. Endolysin shows synergistic action with phage lysin LySMP or antibiotics that are very specific to cell wall components and is considered an alternative to drug antimicrobial therapy because lysine kills target bacteria rather than other microorganisms. There has been a significant increase in potential applications of phage lysin which is specifically promising, kind of topically applied therapeutics, and lysin also becomes a practicable alternative to antibiotics in long-term systemic therapy [117, 142, 143]. Endolysins have been used effectively in medical applications. They exhibit specific antimicrobial activities in controlling and treatment of pathogenic bacteria such as *Streptococcus* and *Staphylococcus*. Beneficial synergistic interactions increase the efficacy of treatments and reduce the risk of resistant strain development. There was no inactivation nor adverse side effects detected *in vivo*. The creation of chimeric proteins by rearrangement of functional domains of lysins of multiple species established molecular engineering of lysins which can increase lytic activity, widen specificity, advance binding affinity, enhance solubility and reduce the chance of resistance formation, thereby optimizing lysins for specific applications. Moreover, endolysin-based antimicrobials viz. (Outer membrane permeabilizers (OMPs) and protein transduction domains (PTDs) are used to control Gram-negative and intracellular pathogens. Molecular engineering of lysins is predicted to gain momentum in the coming years and the dogma of endolysins to be effective only against Gram-positive bacteria when applied externally to decrease [144].

5.9.2 Side effects

Most of the drug resistance in bacteria studies had been carried out on the use of *in vitro* and *in vivo* experiments in animal species with a particular look upon human studies. Predicaments of the misuse of bacteriophages as medicament specialists are in many situations recognized into four classifications: (1) phage selection, (2) bacteriophage have host-range restrictions, (3) the uniqueness of phages as recommended medications, and (4) uncommonness with phage. Studies on T4 phage had recognized no massive fitness effects and pronounced negative results such as inconvenience, itching, wetness, and unattractive scent, unfavourable events, sore throat, belly pain, nausea, extended peristalsis [65, 66]. *Staphylococcus* bacteriophages proved drug intolerance and hypersensitive manifestations at the site of injury on days three to five of bacteriophage therapy, and hepatalgia was detected after several hours [145, 146]. Adverse activities occurred in six (21%) of 28 victims in the Pyophage team in distinction with 13 (41%) of 32 victims in the placebo group Urinary tract infection intravesical Pyobacteriophage (Pyophage;

20 mL) [112]. A cocktail of 12 lytic anti-*P aeruginosa* bacteriophages in the PP1131 group, twenty-three (23%) of 13 analysable individuals had adverse reactions versus seven (54%) of 13 in standard-care-group [83]. It is gratifying to understand renewed interest in bacteriophages which is nature's different tailored solution to the problem of antibiotic-resistant bacteria. Beyond the urgent problem of untreatable infections, detailed research of bacteriophages have the possibility of finding, exhilarating new biology, molecular mechanisms of RNA-guided DNA targeting and cleavage by the Cas9 enzyme Cas9 in genome engineering effective use of CRISPR-mediated understanding of CRISPR–Cas9 mechanisms genome engineering in clinical applications CRISPR biology exemplified via the transformative discovery of CRISPR–Cas DNA editing structures and phage-encoded anti-CRISPR defences [51]. We're on the verge of entering an exciting new era in phage technology and applications, thanks to advanced molecular methods, devices, and applications in medicines.

6. Conclusions

In this chapter, the lytic bacteriophages' efficacy has been reported. Therapeutic bacteriophages have been shown to prevent the growth and replication of a variety of pathogenic bacteria in humans, resulting in improved recovery, health, and survival of infected individuals. Since no or few side effects have been reported, phage therapy is medically safe and effective against bacterial infections. Personalized treatment is presented for phage-resistant gangrene bacterial strains, burn wounds, chronic ulcers, psoriasis, bacterial diarrhoea, urinary tract infections, pneumonia and tuberculosis. Producing higher-quality phage cocktails against specific bacteria groups and making them readily available in all areas, regardless of geography, economics, or climatic conditions, is, however, advantageous. Phage therapy is one of the most effective methods for controlling microbial infections that occur in a variety of species at different times. Expanding research to other organisms may be one of the most useful techniques for collecting evidence and validating the phage therapy's utility and therapeutic potential. Understanding infection mechanisms, phage tolerance, phage therapy effectiveness on targeted pathogens, and their effects on the normal microbiome can all aid in improving biocontrol strategies. A scientific logical approach is needed to develop long term storage and transport of therapeutic bacteriophages with a common guideline for the use and safety of phage therapy. The production of new medicines, innovative methods, and management practises to mitigate the risk of infectious agents being introduced and to reduce predisposing factors may be needed in the future to control bacterial diseases. The discovery of novel phage-host interaction methods and the understanding of how bacteriophages control their hosts will be aided by future studies on the complexities of phage lifestyles and dynamics, bionomics in natural systems, genome and viriome analysis, proteome analysis, genes coding for their proteins, and DNA polymerase phylogeny. To reduce the risk of infectious agents being introduced and to reduce predisposing factors, future bacterial disease control would depend on the development of new drugs, methods, and management practises. In response to the threat posed by multiresistant "super bugs," the use of phage endolysins, as well as possible applications of these enzymes in medicine, food protection, agriculture and veterinary medicine, biotechnology, and environmental sciences, has increased significantly. The significance and trend of research on bacteriophages and their applications is expected to continue as the quest for new antimicrobials intensifies in the near future.

Acknowledgements

The authors thank Ms. Janana Priya for her assistance in the work. The authors also acknowledge support from Sree Balaji Medical College and Hospital for providing facilities and encouragement to complete the work. Govindan Dayanithi belongs to the "Centre National de la Recherche Scientifique-CNRS-French Ministry of Science and Higher Education".

Author's contribution

Both the authors (Palaniappan Ramasamy, Govindan Dayanidhi) have made equal contributions in the conception, design, and execution of the described study and in drafting the manuscript writing and discussion.

Conflict of interest

The authors (PR, GD) declare that there is no financial or conflict of interests.

Author details

Ramasamy Palaniappan[1*] and Govindan Dayanithi[1,2]

1 Director-Research, Research and Development Wing, Central Research Laboratory, Sree Balaji Medical College and Hospital, Bharath Institute of Higher Education (BIHER- Bharath University), Chennai, Tamilnadu, India

2 Molecular Mechanisms in Neurodegenerative Diseases Laboratory, MMDN, University of Montpellier, EPHE-Sorbonne, INSERM, UMR-S1198, Montpellier, France

*Address all correspondence to: researchsbmch@gmail.com; ramasamy_p@hotmail.com

References

[1] Marks T, Sharp R. Review: Bacteriophages and biotechnology: A review. Journal of Chemical Technology and Biotechnology. 2000; 75:6-17 https://doi.org/10.1002/(SICI) 1097-4660(200001)75:1<6::AID-JCTB157>3.0.CO;2-A

[2] Palaniappan Ramasamy Phage Therapy for Control of Bacterial Diseases. Crustacea, 2020, https://doi.org/10.5772/intechopen.88043, Intech Open

[3] D'Hérelle F. Surun microbe invisible antagoniste des Bacillus dysentérique. Comptes rendus de l'Académie des Sciences-

[4] Alain Dublanchet, and Shawna Bourne The epic of phage therapy Can J Infect Dis Med Microbiol. 2007 Jan; 18 (1): 15–18. doi: 10.1155/2007/365761

[5] Chanishvili N. Phage therapy—History from twort and D'Herelle through soviet experience to current approaches. Advances in Virus Research. 2012; 83:4-40. DOI: 10.1016/B978-0-12-394438-2.00001-3

[6] Alexander Sulakvelidze, Zemphira Alavidze, and J. Glenn Morris, Jr. Bacteriophage TherapyAntimicrob Agents Chemother. 2001 Mar; 45(3): 649–659.doi: 10.1128/AAC.45.3.649-659.2001

[7] Krestovnikova, V. A. (1947). Phage treatment and phage prophylactics and their approval in the works of the Soviet researchers. J. Microb. Epidemiol. Immunol. 3, 56–65.

[8] Rob Lavigne, Johan Robben Professor Dr. Richard Bruynoghe Bacteriophage, 2012, 2(1):1-4 ,DOI: 10.4161/bact.20024

[9] Derek M Lin, Britt Koskella, and Henry C Lin Phage therapy: An alternative to antibiotics in the age of multi-drug resistance World J Gastrointest Pharmacol Ther. 2017 Aug 6; 8(3): 162–173. doi: 10.4292/wjgpt. v8. i3.162

[10] Srinivasan P, Ramasamy P. Morphological characterization and biocontrol effects of *Vibrio vulnificus* phages against Vibriosis in the shrimp aquaculture environment. Microbial Pathogenesis. 2017; 111:472-480

[11] Gigante, A., Atterbury, R.J. Veterinary use of bacteriophage therapy in intensively-reared livestock. Virol J 16, 155 (2019). https://doi.org/10.1186/s12985-019-1260-3

[12] Helen J. Jones, Christopher G. Shield, and Benjamin M.C. The Application of Bacteriophage Diagnostics for Bacterial Pathogens in the Agricultural Supply Chain: From Farm-to-Fork Swift.PHAGE. Dec 2020.176

[13] Hasan A. Sohail, Aidan Coffey, Krystyna Debrowska, Irmtraud M. Meyer, Mathias Middelboe, Muhammad Sohail, and Martha R.J. Clokie. Bacteriophages: Emerging Applications in Medicine, Food, and Biotechnology. In: PHAGE.Jun 2020.Vol 1(Issue 2) 75-82.http://doi.org/10.1089/phage.2020.29004.has

[14] Talebi Bezmin Abadi, A., Rizvanov, A.A., Haertlé, T. et al. World Health Organization Report: Current Crisis of Antibiotic Resistance. BioNanoSci. 9, 778–788 (2019). https://doi.org/10.1007/s12668-019-00658-4

[15] World Bank Group "Pulling Together to Beat Superbugs Knowledge and Implementation Gaps in Addressing Antimicrobial Resistance" © 2019 International Bank for Reconstruction and Development/The World Bank 1818 H Street NW, Washington, DC 20433

1Centre Hospitalier Intercommunal de Villeneuve-Saint-Georges, France

[16] Professor Dr. Richard Bruynoghe: a 1951 overview of his bacteriophage research spanning three decades, vol 2, p 1– 4. Taylor & Francis, London, United Kingdom.

[17] Timothy F. Landers, RN, CNP, PhD, a Bevin Cohen, MPH,b Thomas E. Wittum, MS, PhD,c and Elaine L. Larson, RN, PhD, FAAN, CICb A Review of Antibiotic Use in Food Animals: Perspective, Policy, and Potential Public Health Rep. 2012 Jan-Feb; 127(1): 4–22. doi: 10.1177/ 003335491212700103

[18] C. Lee Ventola, MS, The Antibiotic Resistance Crisis Part 1: Causes and Threats, P T. 2015 Apr; 40(4): 277–283. PMCID: PMC4378521 PMID: 25859123

[19] New report calls for urgent action to avert antimicrobial resistance crisis. (2019). https://www.who.int/news/item/29-04-2019-new-report-calls-for-urgent-action-to-avert-antimicrobial-resistance-crisis

[20] O'Neill. 2016. Takling drug-resistant infections globally: Final report and recommendations. The review on Antimicrobial Resistance. 2016. https://amr-review.org/Publications.html

[21] Lillian Brown, Charles Langelier, Michael J. A. Reid, Rachel L. Rutishauser, Luke Strnad, Antimicrobial Resistance: A Call to Action! Clinical Infectious Diseases, Volume 64, Issue 1, 1 January 2017, Pages 106–107, https://doi.org/10.1093/cid/ciw678

[22] C Lee Ventola The antibiotic resistance crisis: part 2: management strategies and new agents P T 2015 May; 40(5):344-52.

[23] Mathias Schmelcher, David M Donovan& Martin J Loessner Bacte riophage endolysins as novel antimicrob ials FUTURE MICROBIOLOGYVOL. 7, NO. 10 REVIEW Published Online:3 Oct 2012 https://doi.org/10.2217/fmb .12.97

[24] Bacteriophage Market Expected To Exhibit Steady Growth During The Forecast Period, Acute Market Reports "Bacteriophage Market - Growth, Future Prospects, Competitive Analysis, 2018 - 2026,". https://www.acutemarketreports.com/report/bacteriophage-market

[25] D H Bergey; John G Holt Bergey's manual of determinative bacteriology. Baltimore: Williams & Wilkins, [1994] ©1994

[26] David R. Boone, Richard W. Castenholz, editors, volume 1-5 ; George M. Garrity, editor-in-chief ; editorial board, James T. Staley ... [et al.] Bergey's manual of systematic bacteriology / Boone, David R.; Castenholz, Richard W.; Garrity, George M.; Bergey, D. H. (David Hendricks), 1860-1937.2001.

[27] Mariateresa Ferone, Aoife Gowen, Séamus Fanning, Amalia G. M. Scannell Microbial detection and identification methods: Bench top assays to omics approaches Comprehensive Reviews in Food Science and Food Safety, 07 September 2020 https://doi.org/10.1111/1541-4337.12618

[28] Børsheim KY. Native marine bacteriophages. FEMS Microbiology Ecology. 1993; 102:141-159

[29] Heldal M, Bratbak G. Production and decay of viruses in aquatic environments. Marine Ecology Progress Series. 1991; 72:205-212

[30] Spichler, A., Hurwitz, B. L., Armstrong, D. G., and Lipsky, B. A. (2015). Microbiology of diabetic foot infections: from Louis Pasteur to 'crime scene investigation'. BMC Med. 13:2. doi: 10.1186/s12916-014-0232-0

[31] Wolcott, R. D., Hanson, J. D., Rees, E. J., Koenig, L. D., Phillips, C. D., Wolcott, R. A., et al. (2016). Analysis of the chronic wound microbiota of 2,963 patients by 16S rDNA pyrosequencing. Wound Repair Regen. 24, 163–174. doi: 10.1111/wrr.12370

[32] Malik, A., Mohammad, Z., and Ahmad, J. (2013). The diabetic foot infections: biofilms and antimicrobial resistance. Diabetes Metab. Syndr. 7, 101–107. doi: 10.1016/j.dsx.2013.02.006

[33] Rahim, K., Qasim, M., Rahman, H., Khan, T. A., Ahmad, I., Khan, N., et al. (2016). Antimicrobial resistance among aerobic biofilm producing bacteria isolated from chronic wounds in the tertiary care hospitals of Peshawar, Pakistan. J. Wound Care 25, 480–486. doi: 10.12968/jowc.2016.25.8.480

[34] Stephanie LaVergne, Theron Hamilton, Biswajit Biswas,Phage Therapy for a Multidrug-Resistant *Acinetobacter baumannii* Craniectomy Site Infection | Open Forum Infectious Diseases | Oxford Academic. Volume 5, Issue 4, April 2018, ofy064

[35] Di Domenico, E. G., Farulla, I., Prignano, G., Gallo, M. T., Vespaziani, M., Cavallo, I., et al. (2017). Biofilm is a major virulence determinant in bacterial colonization of chronic skin ulcers independently from the multidrug resistant phenotype. Int. J. Mol. Sci. 18: E1077. doi: 10.3390/ijms18051077

[36] Markoishvili, K., Tsitlanadze, G., Katsarava, R., Morris, J. G., and Sulakvelidze, A. (2002). A novel sustained-release matrix based on biodegradable poly (ester amide) s and impregnated with bacteriophages and an antibiotic show promise in management of infected venous stasis ulcers and other poorly healing wounds. Int. J. Dermatol. 41, 453–458. doi: 10.1046/j.1365-4362.2002.01451.x

[37] Fish, R., Kutter, E., Wheat, G., Blasdel, B., Kutateladze, M., and Kuhl, S. (2016). Bacteriophage treatment of intransigent diabetic toe ulcers: a case series. J. Wound Care 25, S27–S33. doi: 10.12968/jowc.2016.25.7. S27

[38] Morozova, V. V., Kozlova, Y. u., Ganichev, D., and Tikunova, N. (2018). Bacteriophage treatment of infected diabetic foot ulcers. Methods Mol. Biol. 1693, 151–158. doi: 10.1007/978-1-4939-7395-8_13

[39] Thiel K. Old dogma, new tricks—21st century phage therapy. Nature Biotechnology. 2004;1:31-36

[40] Chan, B. K. et al. Phage selection restores antibiotic sensitivity in MDR *Pseudomonas aeruginosa*. Sci. Rep. 6, 26717; doi: 10.1038/srep26717 (2016).

[41] Tiffany Luong, Ann-Charlott Salabarria, Robert A. Edwards Standardized bacteriophage purification for personalized phage therapy | Nature Protocols. volume 15, pages2867–2890 (2020)

[42] Sarker SA, Sultana S, Reuteler G, Moine D, Descombes P, Charton F, Bourdin G, McCallin S, Ngom-Bru C, Neville T, et al. Oral phage therapy of acute bacterial diarrhea with two coliphage preparations: a randomized trial in children from Bangladesh. EBioMedicine. 2016; 4:124–37.

[43] Jeon, J., Park, J.-H., & Yong, D. (2019). Efficacy of bacteriophage treatment against carbapenem-resistant *Acinetobacter baumannii* in *Galleria mellonella* larvae and a mouse model of acute pneumonia. BMC Microbiology, 19(1), 70

[44] Cao, F., Wang, X., Wang, L., Li, Z., Che, J., Wang, L., Li, X., Cao, Z., Zhang, J., Jin, L., & Xu, Y. (2015, March 24). Evaluation of the Efficacy of a Bacteriophage in the Treatment of Pneumonia Induced by Multidrug Resistance *Klebsiella pneumoniae* in Mice [Research Article]. BioMed Research

International; Hindawi. Volume 2015, Article ID 752930, 9 pages

[45] James M. Regeimbal, Anna C. Jacobs, Brendan W. Corey.Personalized Therapeutic Cocktail of Wild Environmental Phages Rescues Mice from *Acinetobacter baumannii* Wound Infections | Antimicrobial Agents and Chemotherapy. 2016. 02877-15

[46] Shi, Y., Zhao, W., Liu, G., Ali, T., Chen, P., Liu, Y., Kastelic, J. P., Han, B., & Gao, J. (2021). Bacteriophages isolated from dairy farm mitigated *Klebsiella pneumoniae*-induced inflammation in bovine mammary epithelial cells cultured in vitro. BMC Veterinary Research, 17(1), 37.

[47] Gill, J. J., & Hyman, P. (2010). Phage choice, isolation, and preparation for phage therapy. Current Pharmaceutical Biotechnology, 11(1), 2–14. https://doi.org/10.2174/138920110790725311

[48] Smrekar F., Ciringer M., Peterka M., Podgornik A., Strancar A. (2008). Purification and concentration of bacteriophage T4 using monolithic chromatographic supports. J. Chromatogr. B Analyt. Technol. Biomed. Life Sci. 861 177–180. 10.1016/j.jchromb.2007.05.048 [PubMed] [CrossRef] [Google Scholar]

[49] Smrekar F., Ciringer M., Strancar A., Podgornik A. (2011). Characterisation of methacrylate monoliths for bacteriophage purification. J. Chromatogr. 1218 2438–2444. 10.1016/j.chroma.2010.12.083 [PubMed] [CrossRef] [Google Scholar]

[50] Van Belleghem, J. D., Clement, F., Merabishvili, M., Lavigne, R., & Vaneechoutte, M. (2017). Pro- and anti-inflammatory responses of peripheral blood mononuclear cells induced by *Staphylococcus aureus* and *Pseudomonas aeruginosa* phages. Scientific Reports, 7 (1), 8004.

[51] Merabishvili M, Pirnay JP, Verbeken G, Chanishvili N, Tediashvili M, Lashkhi N, Glonti T, Krylov V, Mast J, Van Parys L, et al. Quality-controlled small-scale production of a well-defined bacteriophage cocktail for use in human clinical trials. PLoS ONE. 2009;4(3): e4944.

[52] Amanda Carroll-Portillo, Cristina N. Coffman, Matthew G. Varga, Joe Alcock, Sudha B. Singh, and Henry C. Lin1,4,* Standard Bacteriophage Purification Procedures Cause Loss in Numbers and Activity Viruses. 2021 Feb; 13(2): 328. 2021 Feb 20. doi: 10.3390/v13020328

[53] Bourdin G, Schmitt B, Marvin Guy L, Germond JE, Zuber S, Michot L, Reuteler G, Brussow H. Amplification and purification of T4-like *Escherichia coli* phages for phage therapy: from laboratory to pilot scale. Appl Environ Microbiol. 2013;80(4):1469–76.

[54] Boratynski J, Syper D, Weber-Dabrowska B, Lusiak-Szelachowska M, Pozniak G, Gorski A. Preparation of endotoxin-free bacteriophages. Cell Mol Biol Lett. 2004;9(2):253–9.

[55] Hashemi H, Pouyanfard S, Bandehpour M, Mahmoudi M, Bernasconi M, Kazemi B, Mokhtari-Azad T. Efficient endotoxin removal from T7 phage preparations by a mild detergent treatment followed by ultrafiltration. Acta Virol. 2013;57(3): 373–4.

[56] Cui Z, Feng T, Gu F, Li Q, Dong K, Zhang Y, Zhu Y, Han L, Qin J, Guo X. Characterization and complete genome of the virulent Myoviridae phage JD007 active against a variety of *Staphylococcus aureus* isolates from different hospitals in Shanghai, China. Virol J. 2017;14(1): 26.

[57] Bonilla N, Rojas MI, Netto Flores Cruz G, Hung SH, Rohwer F, Barr JJ.

Phage on tap-a quick and efficient protocol for the preparation of bacteriophage laboratory stocks. PeerJ. 2016;4:e2261.

[58] Srinivasan P, Vaseeharan B, Ramasamy P. Vibrio bacteriophages control the growth of bacterial populations in the aquatic environment. In: The Sixth Indian Fisheries Forum organized by Central Institute of Fisheries Education, Versova, Mumbai —400 061, India; during 17–20 December 2002; 16

[59] Srinivasan P, Ramasamy P. Effect of pH, temperature, enzymes, organic solvents and detergents in the survival and infectivity of *Vibrio bacteriophages*. In: Conference on Microbiology of the Tropical Seas (COMITS) National Institute of Oceanography; Goa, India during 13–15 December 2004

[60] Srinivasan P, Ramasamy P, Brennan GP, et al. Inhibitory effects of bacteriophages on the growth of Vibrio sp. pathogens of shrimp in the Indian aquaculture environment. Asian Journal of Animal and Veterinary. 2007;2(4): 166-183

[61] Stalin N, Srinivasan P. Characterization of *Vibrio parahaemolyticus* and its specific phage from shrimp pond in Palk Strait, South East coast of India. Biologicals. 2016;44 (6):1-8. DOI: 10.1016/j.biologicals.2016.08.003

[62] Jiang SC, Kellogg CA, Paul JH. Characterization of marine temperate phage-host systems isolated from Mamala Bay, Oahu, Hawaii. Applied and Environmental Microbiology. 1998; 64:535-542

[63] Kalatzis PG, Bastías R, Kokkari C, Katharios P. Isolation and characterization of two lytic bacteriophages, φSt2 and φGrn1; phage therapy application for biological control of *Vibrio alginolyticus* in aquaculture live feeds. PLoS One. 2016; 11(3):e0151101. DOI: 10.1371/journal.pone.0151101. eCollection 2016

[64] Adamek Z. Effect of ascogen probiotics supplementation on the growth rate of rainbow trout *Oncorhynchus mykiss* under conditions of intensive culture. Zivocisna Vyroba. 1994;39(3):247-253

[65] Krishnika A, Ramasamy P. Antimicrobial resistance profile of Vibrio species isolated from the hatchery system of *Macrobrachium rosenbergii* (Deman). Indian Journal of Fisheries. 2014;60(4):147-152

[66] Park SC, Shimamura I, Fukunaga M, et al. Isolation of bacteriophages specific to a fish pathogen, *Pseudomonas plecoglossicida*, as a candidate for disease control. Applied and Environmental Microbiology. 2000; 66(4):1416-1422

[67] Srinivasan P, Ramasamy P. Occurrence, distribution and antibiotic resistance patterns of *Vibrio* species associated with viral diseased shrimp of south Indian aquaculture environment. International Journal of Agriculture Sciences. 2009;1(2):1-10

[68] Karunasagar I, Pai R, Malathi GR, et al. Mass mortality of *Penaeus monodon* larvae due to antibiotic-resistant Vibrio harveyi infection. Aquaculture. 1994; 128(3-4):203-209

[69] Jayakumar R, Ramasamy P. Prevalence, biotyping and resistotyping of *Pseudomonas spp.* and *Vibrio sp* isolated from *Penaeus indicus* of Ennore Estuary, Madras, India. In: Chou LM, Munro AD, Lam TJ, Chen TW, Cheong LKK, Ding JK, et al. editors. Proceedings of Third Asian Fisheries Forum, Singapore. 1994. pp. 335-338

[70] Vaseeharan B, Ramasamy P, Murugan T, Chen JC. *In vitro* susceptibility of antibiotics against

Vibrio spp. and *Aeromonas* spp. isolated from *Penaeus monodon* hatcheries and ponds. International Journal of Antimicrobial Agents. 2005; 26:285-291

[71] Ramasamy P, Sujatha Rani J, Gunasekaran DR. Assessment of antibiotic sensitivity and pathogenicity of *Vibrio* spp. and *Aeromonas* spp. from aquaculture environment. MOJ Ecology & Environmental Sciences. 2018;3(3): 128-136

[72] Vaseeharana B, Lin J, Ramasamy P. Effect of probiotics, antibiotic sensitivity, pathogenicity, and plasmid profiles of *Listonella anguillarum*-like bacteria isolated from *Penaeus monodon* culture systems. Aquaculture. 2004;241: 77-91

[73] Międzybrodzki R, Fortuna W, Weber-Dąbrowska B, Górski A. Phage therapy of staphylococcal infections (including MRSA) may be less expensive than antibiotic treatment. Postepy Hig Med Dosw. 2007;61: 461-465

[74] El Haddad L, Harb CP, Gebara MA, Stibich MA, Chemaly RF. A systematic and critical review of bacteriophage therapy against multi-drug resistant ESKAPE organisms in humans. Clin Infect Dis. 2018;69(1):167–78.

[75] Malik DJ, Sokolov IJ, Vinner GK, Mancuso F, Cinquerrui S, Vladisavljevic GT, Clokie MRJ, Garton NJ, Stapley AGF, Kirpichnikova A. Formulation, stabilisation and encapsulation of bacteriophage for phage therapy. Adv Coll Interface Sci. 2017; 249:100–33.

[76] Vandenheuvel D, Singh A, Vandersteegen K, Klumpp J, Lavigne R, Van den Mooter G. Feasibility of spray drying bacteriophages into respirable powders to combat pulmonary bacterial infections. Eur J Pharm Biopharm. 2013; 84(3):578–82.

[77] Golshahi L, Lynch KH, Dennis JJ, Finlay WH. In vitro lung delivery of bacteriophages KS4-M and PhiKZ using dry powder inhalers for treatment of *Burkholderia cepacia* complex and *Pseudomonas aeruginosa* infections in cystic fibrosis. J Appl Microbiol. 2011; 110(1):106–17.

[78] Hoe S, Semler DD, Goudie AD, Lynch KH, Matinkhoo S, Finlay WH, Dennis JJ, Vehring R. Respirable bacteriophages for the treatment of bacterial lung infections. J Aerosol Med Pulm Drug Del. 2013;26(6):317–35.

[79] Josef Prazak; Luca Valente; Manuela Iten; Lea Federer, Denis Grandgirard; Sara Soto; Gregory Resch; Stephen L. Leib; Stephan M. Jakob; Matthias Haenggi; David R. Cameron; Yok-Ai Que. Benefits of aerosolized phages for the treatment of pneumonia due to methicillin-resistant *Staphylococcus aureus* (MRSA): an experimental study in rats. 2021 The Journal of Infectious Diseases, jiab112, https://doi.org/10.1093/infdis/jiab112

[80] Schooley RT, B, Gill JJ, Hernandez-Morales A, Lancaster J, Lessor L, Barr JJ, Reed SL, Rohwer F, Benler S, et al. Development and use of personalized bacteriophage-based therapeutic cocktails to treat a patient with a disseminated resistant *Acinetobacter baumannii* infection. Antimicrob Agents Chemother. 2017;61(10):e00954.

[81] Cano, E. J., Caflisch, K. M., Bollyky, P. L., Van Belleghem, J. D., Patel, R., Fackler, J., Brownstein, M. J., Horne, B., Biswas, B., Henry, M., Malagon, F., Lewallen, D. G., & Suh, G. A. (2020). Phage Therapy for Limb-threatening Prosthetic Knee *Klebsiella pneumoniae* Infection: Case Report and In Vitro Characterization of Anti-biofilm Activity. Clinical Infectious Diseases, ciaa705.

[82] Colom J, Cano-Sarabia M, Otero J, Arinez-Soriano J, Cortes P, Maspoch D,

Llagostera M. Microencapsulation with alginate/CaCO3: a strategy for improved phage therapy. Sci Rep. 2017; 7:41441.

[83] Jault P, Leclerc T, Jennes S, Pirnay JP, Que YA, Resch G, Rousseau AF, Ravat F, Carsin H, Le Floch R, et al. Efficacy and tolerability of a cocktail of bacteriophages to treat burn wounds infected by *Pseudomonas aeruginosa* (PhagoBurn): a randomised, controlled, double-blind phase 1/2 trial. Lancet Infect Dis. 2018;19(1):35–45.

[84] Boucher HW, Talbot GH, Bradley JS, Edwards JE, Gilbert D, Rice LB, Scheld M, Spellberg B, Bartlett J. Bad bugs, no drugs: no ESKAPE! an update from the infectious disease's society of America. Clin Infect Dis. 2009;48(1):1–12.

[85] Habusha M, Tzipilevich E, Fiyaksel O, Ben-Yehuda S. A mutant bacteriophage evolved to infect resistant bacteria gained a broader host range. Mol Microbiol. 2019;111(6):1463–75.

[86] Gu J, Liu X, Li Y, Han W, Lei L, Yang Y, Zhao H, Gao Y, Song J, Lu R, et al. A method for generation phage cocktail with great therapeutic potential. PLoS ONE. 2012;7(3):e31698

[87] Mathias Schmelcher, David M Donovan& Martin J Loessner Bacte riophage endolysins as novel antimicrob ials FUTURE MICROBIOLOGYVOL. 7, NO. 10 REVIEW Published Online:3 Oct 2012https://doi.org/10.2217/fmb.12.97

[88] Cha, K., Oh, H. K., Jang, J. Y., Jo, Y., Kim, W. K., Ha, G. U., Ko, K. S., & Myung, H. (2018). Characterization of Two Novel Bacteriophages Infecting Multidrug-Resistant (MDR) *Acinetobacter baumannii* and Evaluation of Their Therapeutic Efficacy in Vivo. Frontiers in Microbiology, 9-9:696.

[89] Frank Oechslin, Philippe Piccardi, Stefano Mancini.Synergistic Interaction Between Phage Therapy and Antibiotics Clears *Pseudomonas aeruginosa* Infection in Endocarditis and Reduces Virulence | The Journal of Infectious Diseases | Oxford Academic. Volume 215, Issue 5, 1 March 2017, Pages 703–712

[90] Heo, Y.-J., Lee, Y.-R., Jung, H.-H., Lee, J., Ko, G., & Cho, Y.-H. (2009). Antibacterial Efficacy of Phages against *Pseudomonas aeruginosa* Infections in Mice and *Drosophila melanogaster*. Antimicrobial Agents and Chemotherapy, 53(6), 2469–2474.

[91] Rosanna Capparelli, Marianna Parlato, Giorgia Borriello.Experimental Phage Therapy against Staphylococcus aureus in Mice | Antimicrobial Agents and Chemotherapy.2007. 01513-06

[92] Sivera Marza, J. A., Soothill, J. S., and Boydell, P. (2006). Multiplication of therapeutically administered bacteriophages in *Pseudomonas aeruginosa* infected patients. Burns 32, 644–646. doi: 10.1016/j.burns.2006.02.012

[93] Cerveny, K. E., DePaola, A., Duckworth, D. H., & Gulig, P. A. (2002). Phage Therapy of Local and Systemic Disease Caused by Vibrio vulnificus in Iron-Dextran-Treated Mice. Infection and Immunity, 70(11), 6251–6262.

[94] Biswajit Biswas, Sankar Adhya, Paul Washart.Bacteriophage Therapy Rescues Mice Bacteremic from a Clinical Isolate of Vancomycin-Resistant *Enterococcus faecium* | Infection and Immunity. .70.1.204-210.2002 (n.d.).

[95] Scott LaFee, Heather Buschman. Novel Phage Therapy Saves Patient with Multidrug-Resistant Bacterial Infection. (n.d.). UC Health - UC San Diego. 2017

[96] Vlassov, V. V., Ganichev, D. A., Kozlova, J. N., Morozova, V. V., Saranina, I. V., and Tikunova, N. V. (2016). "Personalised phage therapy of infected trophic ulcers on the

background of diabetis," in Abstract Retrieved from Book of Abstracts of 3-rd International Scientific Conference Bacteriophages: Theoretical and Practical Aspects of Their Application in Medicine, Veterinary and Food. Available online at: http://www.congress-phages.ru/_pictures/tezis_bf-2016_block.pdf

[97] Chan, B. K., Turner, P. E., Kim, S., Mojibian, H. R., Elefteriades, J. A., & Narayan, D. (2018). Phage treatment of an aortic graft infected with *Pseudomonas aeruginosa*. Evolution, Medicine, and Public Health, 2018(1), 60–66.

[98] C Duplessis, B Biswas, B Hanisch. Refractory *Pseudomonas* Bacteremia in a 2-Year-Old Sterilized by Bacteriophage Therapy | Journal of the Pediatric Infectious Diseases Society | Oxford Academic. Volume 7, Issue 3, September 2018, Pages 253–256

[99] Ho K, Huo W, Pas S, Dao R, Palmer KL. Loss of function mutations in epaR confer resistance to phage NPV1 infection in *Enterococcus faecalis* OG1RF. Antimicrob Agents Chemother. 2018;62 (10):e00758

[100] Zhang QG, Buckling A. Phages limit the evolution of bacterial antibiotic resistance in experimental microcosms. Evol Appl. 2012;5(6):575–82

[101] Kochetkova, V. A., Mamontov, A. S., Moskovtseva, R. L., Erastova, E. I., Trofimov, E. I., Popov, M. I., et al. (1989). Phagotherapy of postoperative suppurative-inflammatory complications in patients with neoplasms. Sov. Med. 6, 23–26.

[102] Zhukov-Verezhnikov, N. N., Peremitina, L. D., Berillo, E. A., Komissarov, V. P., and Bardymov, V. M. (1978). Therapeutic effect of bacteriophage preparations in the complex treatments of suppurative surgical diseases. Sov. Med. 12, 64–66.

[103] Lazareva, E. B., Smirnov, S. V., Khvatov, V. B., Spiridonova, T. G., Bitkova, E. E., Darbeeva, O. S., et al. (2001). Efficacy of bacteriophages in complex treatment of patients with burn wounds. Antibiot. Khimioter. 46, 10–14.

[104] Erol, S., Altoparlak, U., Akcay, M. N., Celebi, F., and Parlak, M. (2004). Changes of microbial flora and wound colonization in burned patients. Burns 4, 357–361. doi: 10.1016/j.burns.2003.12.013

[105] Church, D., Elsayed, S., Reid, O., Winston, B., and Lindsay, R. (2006). Burn wound infections. Clin. Microbiol. Rev. 19, 403–434. doi: 10.1128/CMR.19.2.403-434.2006

[106] Rose, T., Verbeken, G., Vos, D. D., Merabishvili, M., Vaneechoutte, M., Lavigne, R., et al. (2014). Experimental phage therapy of burn wound infection: difficult first steps. Int. J. Burns Trauma 4, 66–73.

[107] Asati, S., and Chaudhary, U. (2017). Prevalence of biofilm producing aerobic bacterial isolates in burn wound infections at a tertiary care hospital in northern India. Ann. Burns Fire Disasters 30, 39–42.

[108] McVay, C. S., Velásquez, M., & Fralick, J. A. (2007). Phage Therapy of *Pseudomonas aeruginosa* Infection in a Mouse Burn Wound Model. Antimicrobial Agents and Chemotherapy, 51(6), 1934–1938

[109] Soothill, J. S. (1994). Bacteriophage prevents destruction of skin grafts by Pseudomonas aeruginosa. Burns 20, 209–211. doi: 10.1016/0305-4179(94) 90184-8

[110] Rhoads, D. D., Wolcott, R. D., Sun, Y., and Dowd, S. E. (2012). Comparison of culture and molecular identification of bacteria in chronic wounds. Int. J.

Mol. Sci. 13, 2535–2550. doi: 10.3390/ijms13032535

[111] Kvachadze, L., Balarjishvili, N., Meskhi, T., Tevdoradze, E., Skhirtladze, N., and Pataridze, T. (2011). Evaluation of lytic activity of staphylococcal bacteriophage Sb-1 against freshly isolated clinical pathogens. Microb. Biotechnol. 4, 643–650. doi: 10.1111/j.1751-7915.2011. 00259.x

[112] Lorenz Leitner, MD., Aleksandre Ujmajuridze, MD. Intravesical bacteriophages for treating urinary tract infections in patients undergoing transurethral resection of the prostate: A randomised, placebo-controlled, double-blind clinical trial—The Lancet Infectious Diseases. 2020; 20: 1263–72

[113] Ujmajuridze, A., Chanishvili, N., Goderdzishvili, M., Leitner, L., Mehnert, U., Chkhotua, A., Kessler, T. M., & Sybesma, W. (2018). Adapted Bacteriophages for Treating Urinary Tract Infections. Frontiers in Microbiology, 9.

[114] Khawaldeh, A., Morales, S., Dillon, B., Alavidze, Z., Ginn, A. N., Thomas, L., Chapman, S. J., Dublanchet, A., Smithyman, A., & Iredell, J. R. (2011). Bacteriophage therapy for refractory *Pseudomonas aeruginosa* urinary tract infection. Journal of Medical Microbiology, 60(11), 1697–1700.

[115] Heselpoth, R. D., Euler, C. W., Schuch, R., & Fischetti, V. A. (2019). Lysocins: Bioengineered Antimicrobials That Deliver Lysins across the Outer Membrane of Gram-Negative Bacteria. Antimicrobial Agents and Chemotherapy, 63(6). https://doi.org/10.1128/AAC.00342-19

[116] Dedrick RM, Guerrero-Bustamante CA, Garlena RA, Russell DA, Ford K, Harris K, Gilmour KC, Soothill J, Jacobs-Sera D, Schooley RT, et al. Engineered bacteriophages for treatment of a patient with a disseminated drug-resistant *Mycobacterium abscessus*. Nat Med. 2019;25(5):730–3.

[117] Witzenrath, M., Schmeck, B., Doehn, J. M., Tschernig, T., Zahlten, J., Loeffler, J. M., Zemlin, M., Müller, H., Gutbier, B., Schütte, H., Hippenstiel, S., Fischetti, V. A., Suttorp, N., & Rosseau, S. (2009). Systemic use of the endolysin Cpl-1 rescues mice with fatal *Pneumococcal pneumonia*. Critical Care Medicine, 37(2), 642–649.

[118] WHO | WHO consolidated guidelines on drug-resistant tuberculosis treatment. (2019). WHO; World Health Organization? ISBN 978-92-4-155052-9. http://www.who.int/tb/publications/2019/consolidated-guidelines-drug-resistant-TB-treatment/en/

[119] Smith, I. (2003). *Mycobacterium tuberculosis* Pathogenesis and Molecular Determinants of Virulence. Clinical Microbiology Reviews, 16(3), 463–496. https://doi.org/10.1128/CMR.16.3.463-496.2003

[120] Walsh, D. S., Portaels, F., & Meyers, W. M. (2010). Recent advances in leprosy and Buruli ulcer (*Mycobacterium ulcerans* infection). Current Opinion in Infectious Diseases, 23(5), 445–455. https://doi.org/10.1097/QCO.0b013e32833c2209

[121] Hatfull, G. F. (2018). Mycobacteriophages. Microbiology Spectrum, 6(5). https://doi.org/10.1128/microbiolspec.GPP3-0026-2018

[122] Joseph Antony Sundarsingh.T, Ranjitha.J, Features of the biochemistry of *Mycobacterium smegmatis*, as a possible model for *Mycobacterium tuberculosis*—ScienceDirect. (2020). Volume 13, Issue 9, Pages 1255-1264 https://www.sciencedirect.com/science/article/pii/S1876034120305530

[123] Carrigy, N. B., Larsen, S. E., Reese, V., Pecor, T., Harrison, M., Kuehl, P. J.,

Hatfull, G. F., Sauvageau, D., Baldwin, S. L., Finlay, W. H., Coler, R. N., & Vehring, R. (2019). Prophylaxis of *Mycobacterium tuberculosis* H37Rv Infection in a Preclinical Mouse Model via Inhalation of Nebulized Bacteriophage D29. Antimicrobial Agents and Chemotherapy, 63(12). https://doi.org/10.1128/AAC.00871-19

[124] Liu, K., Yang, W., Dong, X., Cong, L., Li, N., Li, Y., Wen, Z., Yin, Z., Lan, Z., Li, W., & Li, J. (2016). Inhalation Study of Mycobacteriophage D29 Aerosol for Mice by Endotracheal Route and Nose-Only Exposure. Journal of Aerosol Medicine and Pulmonary Drug Delivery, 29. https://doi.org/10.1089/jamp.2015.1233

[125] Trigo, G., Martins, T. G., Fraga, A. G., Longatto-Filho, A., Castro, A. G., Azeredo, J., & Pedrosa, J. (2013). Phage therapy is effective against infection by *Mycobacterium ulcerans* in a murine footpad model. PLoS Neglected Tropical Diseases, 7(4), e2183. https://doi.org/10.1371/journal.pntd.0002183

[126] Danelishvili, L., Young, L., & Luiz, E. (2006). In Vivo Efficacy of Phage Therapy for *Mycobacterium avium* Infection As Delivered by a Nonvirulent Mycobacterium. Microbial Drug Resistance (Larchmont, N.Y.), 12, 1–6. https://doi.org/10.1089/mdr.2006.12.1

[127] Latka, A., & Drulis-Kawa, Z. (2020). Advantages and limitations of microtiter biofilm assays in the model of antibiofilm activity of *Klebsiella* phage KP34 and its depolymerase. Scientific Reports, 10(1), 20338. https://doi.org/10.1038/s41598-020-77198-5

[128] Liu, Y., Leung, S. S. Y., Huang, Y., Guo, Y., Jiang, N., Li, P., Chen, J., Wang, R., Bai, C., Mi, Z., & Gao, Z. (2020). Identification of Two Depolymerases From Phage IME205 and Their Antivirulent Functions on K47 Capsule of *Klebsiella pneumoniae*. Frontiers in Microbiology, 11. https://doi.org/10.3389/fmicb.2020.00218

[129] Kim, S., Lee, D.-W., Jin, J.-S., & Kim, J. (2020). Antimicrobial activity of LysSS, a novel phage endolysin, against *Acinetobacter baumannii* and *Pseudomonas aeruginosa*. Journal of Global Antimicrobial Resistance, 22, 32–39. https://doi.org/10.1016/j.jgar.2020.01.005

[130] Becker, S. C., Roach, D. R., Chauhan, V. S., Shen, Y., Foster-Frey, J., Powell, A. M., Bauchan, G., Lease, R. A., Mohammadi, H., Harty, W. J., Simmons, C., Schmelcher, M., Camp, M., Dong, S., Baker, J. R., Sheen, T. R., Doran, K. S., Pritchard, D. G., Almeida, R. A., ... Donovan, D. M. (2016). Triple-acting Lytic Enzyme Treatment of Drug-Resistant and Intracellular *Staphylococcus aureus*. Scientific Reports, 6(1), 25063. https://doi.org/10.1038/srep25063

[131] Shen, Y., Barros, M., Vennemann, T., Gallagher, D. T., Yin, Y., Linden, S. B., Heselpoth, R. D., Spencer, D. J., Donovan, D. M., Moult, J., Fischetti, V. A., Heinrich, F., Lösche, M., & Nelson, D. C. (2016). A bacteriophage endolysin that eliminates intracellular streptococci. ELife, 5, e13152. https://doi.org/10.7554/eLife.13152

[132] Yang, H., Zhang, H., Wang, J., Yu, J., & Wei, H. (2017). A novel chimeric lysin with robust antibacterial activity against planktonic and biofilm methicillin-resistant *Staphylococcus aureus*. Scientific Reports, 7(1), 40182.

[133] Pastagia, M., Euler, C., Chahales, P., Fuentes-Duculan, J., Krueger, J. G., & Fischetti, V. A. (2011). A Novel Chimeric Lysin Shows Superiority to Mupirocin for Skin Decolonization of Methicillin-Resistant and -Sensitive *Staphylococcus aureus* Strains. Antimicrobial Agents and Chemotherapy, 55(2), 738–744.

[134] Daniel, A., Euler, C., Collin, M., Chahales, P., Gorelick, K. J., & Fischetti, V. A. (2010). Synergism between a Novel Chimeric Lysin and Oxacillin Protects against Infection by Methicillin-Resistant *Staphylococcus aureus*. Antimicrobial Agents and Chemotherapy, 54(4), 1603–1612.

[135] García, P., Martínez, B., Rodríguez, L., & Rodríguez, A. (2010). Synergy between the phage endolysin LysH5 and nisin to kill *Staphylococcus aureus* in pasteurized milk. International Journal of Food Microbiology, 141(3), 151–155.

[136] Celia, L. K., Nelson, D., & Kerr, D. E. (2008). Characterization of a bacteriophage lysin (Ply700) from *Streptococcus uberis*. Veterinary Microbiology, 130(1), 107–117.

[137] Cheng, Q., Nelson, D., Zhu, S., & Fischetti, V. A. (2005). Removal of group B streptococci colonizing the vagina and oropharynx of mice with a bacteriophage lytic enzyme. Antimicrobial Agents and Chemotherapy, 49(1), 111–117.

[138] Yoong, P., Schuch, R., Nelson, D., & Fischetti, V. A. (2004). Identification of a broadly active phage lytic enzyme with lethal activity against antibiotic-resistant *Enterococcus faecalis* and *Enterococcus faecium*. Journal of Bacteriology, 186(14), 4808–4812.

[139] Nelson, D., Loomis, L., & Fischetti, V. A. (2001). Prevention and elimination of upper respiratory colonization of mice by group A streptococci by using a bacteriophage lytic enzyme. Proceedings of the National Academy of Sciences of the United States of America, 98(7), 4107–4112.

[140] Jingmin Gu, Hengyu Xi, Mengjun Cheng & Wenyu Han Phage-derived lysins as therapeutic agents against multidrug-resistant *Enterococcus faecalis* 1FUTURE MICROBIOLOGYVOL. 13, NO. 3EDITORIAL, 2018https://doi.org/10.2217/fmb-2017-0235

[141] Schmelcher M, Powell AM, Becker SC, Camp MJ, Donovan DM. Chimeric phage lysins act synergistically with lysostaphin to kill mastitis-causing *Staphylococcus aureus* in murine mammary glands. Appl. Environ. Microbiol. 2012;78(7):2297–2305. [PMC free article] [PubMed] [Google Scholar]

[142] Haddad Kashani H, Fahimi H, Goli YD, Moniri R. 2017. A novel chimeric endolysin with antibacterial activity against methicillin-resistant *Staphylococcus aureus*. Front Cell Infect Microbiol 7:290. doi:10.3389/fcimb.2017.00290. 105.

[143] O'Flaherty S, Ross RP, Coffey A. Bacteriophage and their lysins for elimination of infectious bacteria. FEMS Microbiology Reviews. 2009;33(4): 801-819

[144] Domenech M, Garcia E, Moscoso M. *In vitro* destruction of *Streptococcus pneumoniae* biofilms with bacterial and phage peptidoglycan hydrolases. Antimicrob. Agents Chemother. 2011;55(9):4144–4148. [PMC free article] [PubMed] [Google Scholar]

[145] Bruttin A, Brussow H. Human volunteers receiving *Escherichia coli* phage T4 orally: a safety test of phage therapy. Antimicrob Agents Chemother. 2005;49(7):2874–8

[146] Wright, A., Hawkins, C. H., Änggård, E. E., & Harper, D. R. (2009). A controlled clinical trial of a therapeutic bacteriophage preparation in chronic otitis due to antibiotic-resistant *Pseudomonas aeruginosa*; a preliminary report of efficacy. Clinical Otolaryngology, 34(4), 349–357. https://doi.org/10.1111/j.1749-4486.2009.01973.x

Chapter 9

Potential of Inhaled Bacteriophage Therapy for Bacterial Lung Infection

Wei Yan, Subhankar Mukhopadhyay,
Kenneth Kin Wah To and Sharon Shui Yee Leung

Abstract

Phage therapy as a promising alternative antimicrobial to treat multidrug resistant (MDR) bacteria related lung infections, has drawn significant attention in clinical trials and bench-scale study in the recent decade, and the therapeutic effect of local delivery of phage has been demonstrated by several clinical reports. This book chapter discusses the current clinical development of inhaled phage therapy followed by the advancement of phage formulation designs for respiratory delivery of phage using various inhalation devices and their *in vivo* efficacy. The development of combination therapy of phage and antibiotics to combat MDR bacteria associated lung infections is also covered to reflect the current clinical practice. Lastly, we also share our insights on the challenges of advancing inhaled phage therapy and potential directions for future research.

Keywords: pulmonary delivery, multidrug-resistant bacteria, respiratory infection, dry powder inhaler, nebulization, phage formulation, inhaled phage therapy

1. Introduction

Lung infection is a leading cause of morbidity and mortality worldwide [1]. Currently, antibiotics remain the mainstay treatment options for bacterial lung infections [2]. With the rapid emergence of multidrug-resistant (MDR) bacteria, last-line antibiotics such as colistin and carbapenem have been increasingly used for life-threatening infections. However, nosocomial outbreaks caused by pan-drug resistant (PDR) 'superbugs' have also been increasingly reported worldwide, creating significant therapeutic challenges for the treatment of lung infections [3–5].

Bacteriophage (phage) therapy has been proposed as a promising alternative to antibiotics in combating bacterial infections, including those caused by the MDR pathogens. A comprehensive review from Abedon summarized earlier clinical studies of phage application, with most reported cases from Eastern Europe as these countries more practical experience [6]. Overall, phage therapy for respiratory infections have not been extensively studied and only a handful of human studies reported [6–8].

Although recent failure of the "Phagoburn" trial against burn wound infections is discouraging, a lesson we learnt is the importance of the stability of phage preparations and the efficient delivery of sufficient amount of viable phage to the site of infections [9]. Pulmonary delivery of phage would hold the greatest promise in achieving optimal concentration of phage in the lung for effective treatment. In this book chapter, we first introduce the clinical progress of inhaled phage therapy and highlight recent advancement made in the delivery of phage preparations using various inhalation devices. As most experimental phage therapeutic investigations were conducted with concomitant antibiotic treatment, we also discuss the development of phage-antibiotic combinations to treat lung infections. Lastly, we summarize the challenges that must be overcome in order to translate inhaled phage therapy to clinical applications.

2. Clinical development of inhaled phage therapy

In the past decade, a few success stories in experimental inhaled phage therapy were reported. Hoyle et al. reported a successful inhaled phage therapy to manage chronic lung infection caused by MDR *Achromobacter xylosoxidans* [10]. The 17-year-old female patient was unsuccessfully treated with many rounds of antibiotics before she was given a phage cocktail treatment containing two *Achromobacter* phages in the Eliava Phage Therapy Center. The phage cocktail was given by nebulization once daily and orally twice daily for 20 days. The treatment was repeated 4 times at 1, 3, 6 and 12 months after the initial treatment. The patient's subjective conditions were significantly improved and her lung function-FEV1 increased from 1.83 L to 3.33 L together with intermittent antibiotic regimen. Successful phage treatment was also reported for a 12-year-old lung-transplanted cystic fibrosis (CF) patient suffered from persistent lung infection caused by PDR *A. xylosoxidans* [11]. After two rounds of inhaled phage therapy, the patient's respiratory condition slowly improved and the bacterial load was significantly reduced. Similar favorable therapeutic efficacy was also reported in another clinical case [12], where a five-year-old cystic fibrosis patient suffering from severe lung infections was treated with a commercially available phage preparation (pyophage) by nebulization.

Aslam et al. reported the early clinical experience of phage therapy in lung transplant recipients in the USA [13]. Three patients with life-threatening MDR infections caused by *Pseudomonas aeruginosa* (n = 2) and *Burkholderia dolosa* (n = 1) received phage cocktails via both intravenous injection and nebulization with concurrent antibiotic treatments for variable duration. Two patients responded clinically with the phage treatments and were discharged from hospitals, while the third patient infected by *B. dolosa* was dead due to infection relapsed. Nonetheless, no phage therapy-related adverse events were identified. While these experimental use of inhaled phage therapy as an adjunct treatment has demonstrated the clinical benefits in treating lung infections caused by MDR superbugs, well-designed clinical trials are needed to convincingly evaluate its clinical efficacy.

To date, there have been three phage therapy clinical studies registered with the ClinicalTrials.gov to evaluate the safety and efficacy of phage therapy against lung infections (**Table 1**). "MUCOPHAGES" (NCT01818206) assessed the effect of a cocktail of 10 phages on *P. aeruginosa* from sputum samples isolated from CF patients. Although the trial was completed in 2012 according to the clinical trial registry, no information about the outcome of this trial was published. In 2020, two other trials were launched. One trial (NCT04636554) is attempting to apply personalized phage treatment in Covid-19 patients with bacterial co-infections microbial for pneumonia or bacteremia/septicemia. Another trial launched by Armata Pharmaceuticals is a Phase 1b/2a, double-blind, randomized, placebo-controlled trial (NCT04596319) aiming to study the safety, tolerability, and preliminary efficacy of inhaled AP-PA02 in subjects with CF and chronic pulmonary

ClinicalTrails.gov Identifier	Phase	Target condition/ Disease	Phage	Design	Trail status
NCT01818206	NA	Cystic Fibrosis	A cocktail of 10 bacteriophages	Single Group Assignment	Completed in 2012
NCT04596319	1b/2a	Chronic *Pseudomonas aeruginosa* Lung Infections and Cystic Fibrosis	AP-PA02 cocktail	Parallel Assignment (Randomized, double-blind, placebo-controlled)	Recruiting as of the preparation of this book chapter
NCT04636554	NA	Covid-19 patients with bacterial co-infections	Phages against *A. baumannii*, *P. aeruginosa* or *S. aureus*	Expanded Access (Intermediate-size Population, Treatment IND/ Protocol)	Recruiting as of the preparation of this book chapter

NA: not available.

Table 1.
Clinical trials of phage therapy for lung infections.

P. aeruginosa infection. This is the first randomized trial on inhaled phage therapy and the AP-PA02 cocktail is an advanced version of AP-PA01 which was used in the successful experimental study documented in Aslam et al. [13]. The findings from this trial are expected to set a landmark for the development of inhaled phage therapy.

3. Nebulization

3.1 Liquid formulation

Majority of the phage studies for lung delivery focus on liquid formulations as minimal formulation development is required to prepare phage cocktails with sufficient stability for a short storage period. The long term storage stability of phage in liquid formulations was often reported. Cooper et al. demonstrated a phage cocktail of 3 *Pseudomonas* phages (GL-1, GL-12.5 and LP-M10) suspended in phosphate buffered saline (PBS) was stable at both 4 °C and room temperature with no statistically significant titer loss ($\leq$ 0.5 log) for 6 months [14]. As most commonly used phage stabilizers, including PBS, salt-magnesium buffer (SMB) and Tris-H buffer are not yet approved for inhalation. Dilution of phage suspension with 0.9% sodium chloride (NaCl) is usually needed for pulmonary administration [10]. Carrigy et al. showed minimal impacts on the phage stability with the NaCl dilution process, suggesting the suitability of this approach [15].

To date, nebulization has been the exclusive choice for pulmonary delivery of phage suspension in human studies due to its high delivery efficiency and capability of delivering a large volume of liquid phage formulation (> 1 mL) to patients including those cannot administer the dose voluntarily. Several types of commercial nebulizers are available to aerosolize phage into fine droplets using different aerosol generation mechanisms, including air-jet nebulization, vibrating mesh nebulization, ultrasonic nebulization, and colliding liquid jets [16, 17]. The suitability of these nebulizers in delivering phage to lungs has been previously evaluated in terms of deactivation of phage upon the nebulization process.

Jet nebulizers use compressed air to atomize the liquid phage suspension into primary droplets and their subsequent impaction onto the baffle would further breakdown into smaller droplets suitable for inhalation. LC-star nebulizer [16, 18], Collison 6-jet [19–21], LC Sprint jet nebulizer [22], AeroEclipse [23] and atomizer [24] have been used to deliver therapeutic phages. Leung et al. showed the air-jet nebulization had negligible impacts on the stability of the *Podovidae* PEV2 phage, while significant titer loss was found in *Myoviridae* PEV40 phage (~1 log loss) and *Siphoviridae* D29 phage (~3 log loss) [22]. Based on the cryo-transmission electron microscopy analysis, they found the nebulization-induced titer loss was correlated with morphological damage to phages. They further suggested that the length of phage tail may be an important consideration when delivering phages via jet nebulization, particularly for phage cocktails containing phages of different morphologies. The influence of the final formulation composition for nebulization of D29 phage was evaluated by Liu et al. using a Collison 6-jet nebulizer [19, 21]. They reported that deionized water was the optimal spray liquid for D29 aerosol generation and they postulated that the high ion strength and salt concentrations in the PBS and 0.9% NaCl were detrimental to the phage upon jet nebulization. These results were in accord with Carrigy et al. and Leung et al. nebulizing buffered D29 using other jet nebulizers [15, 22]. Liu et al. also studied the impact of relative humidity (RH) on the stability of nebulized D29 and found a low environmental humidity condition was more favorable for D29 nebulization [19]. Later, Verreault et al. reported that the stability of nebulized phage aerosols at different temperatures and humidity is phage-dependent with some being more robust and some being more vulnerable [21]. Overall, these studies highlighted the importance of controlling the temperature and RH for phage nebulization.

Vibrating mesh nebulizers produce aerosol droplets by extruding the liquid formulation through a membrane with calibrated holes based on the converse piezoelectric effects. Several studies compared the aerosol delivery of phage between jet and mesh nebulizers [15, 16, 23–25]. Golshahi et al. showed both the LCstar (air-jet) and eFlow (mesh) nebulizers were suitable for the delivery of phages active against *Burkholderia cepacia* Complex by imaging the lung deposition and mathematical model prediction [16]. In some studies, mesh nebulizers were found to be more detrimental to phage than air-jet nebulizers [23, 24], but reasons for the poorer delivery of mesh nebulizer were unclear. In contrast, better phage recovery was noted after nebulizing using a mesh nebulizer compared with the jet nebulization in some other studies [15, 25]. Visual evidence on the correlation between the titer reduction and morphological change of a *Myoviridae* PEV44 phage after nebulization was provided by Leung et al., showing more "intact" phage was detected in the mesh-nebulized phage samples under TEM image. The more destructive effect of jet nebulization is likely caused by stresses associated with the droplet production and re-nebulization processes. Based on the collected experimental data and a mathematical model, Carrigy et al. estimated phage were re-nebulized an average of 96 times before exiting the mouthpiece of the jet nebulizer [15]. A review from Prichard et al. revealed that 86% of the disclosed nebulizer technology have chosen vibration-mesh nebulizers as the delivery devices, particularly for stress-sensitive drugs [26]. The mixed findings of phage nebulization in the literature can be attributed to many factors, such as phage types, formulation composition, experimental conditions (like temperature, humidity and sample collection methods) and different models of the same nebulizer type. Therefore, the survival of individual phages within a cocktail should be tested with different delivery devices for the optimization of phage cocktail – inhalation device combinations.

Ultrasonic nebulizers use a piezoelectric transducer to generate ultrasonic wave in the liquid drug formulation and aerosolize it at the solution surface. Upon the nebulization process, a portion of the ultrasonic energy converts to heat, which could be detrimental to heat-sensitive biologics, like phages. Only one study reported the use of an ultrasonic nebulizer to deliver phage to treat lung infections in a mink model, but little data on the nebulization process was available [27]. More recently, Marqus et al. assessed the capability of a novel low cost and portable hybrid surface and bulk acoustic wave (HYDRA) nebulizer to deliver a *Myoviridae* phage K and lysostaphin to target *Staphylococcus aureus* [28]. Negligible titer reduction was noted (0.1 log loss), possibly due to the relatively low powers and high frequencies (approximately 10 MHz) of the nebulizer. Furthermore, the size of the aerosols generated by HYDRA is smaller (DV50 1.85 μm), well within the respirable range, demonstrating its suitability for pulmonary delivery of phages.

3.2 *In vivo* efficacy of inhaled phage therapy achieved with nebulization

The *in vivo* efficacy of phage liquid formulation has been studied in rodent and mink models. Semler et al. established *B. cenocepacia* respiratory infection model in mice and then treated with liquid phage formulation delivered by a LC-star jet nebulizer or intraperitoneal injection (IP) [18]. After a 2-day treatment, the lung bacterial load was only reduced by ~0.5 log in mice received phage via IP injection, but a 2-log bacterial reduction was observed in mice treated with inhaled phage. This finding is in contradiction with a previous study showing that phage delivered by the IP route was more efficacious than intranasal instillation in treating a *B. cenocepacia* respiratory infection in mice [29]. Semler et al. accounted the discrepancy to the efficiency of phage delivery to lungs that nebulization is a more effective way in delivering phage particles to the lung than intranasal instillation. Also, the capability of IP injected phage reaching lung is significantly affected by the clearance rate of phage in blood which is phage-dependent. The *in vivo* delivery efficiency of D29 phage using a Collison 6-jet nebulizer and IP route was compared by Liu et al. [20]. Approximately 10% of D29 phage could reach to the lung of mice after nebulization and complete phage elimination was noted in 72 h, whereas only 0.1% of the phage could reach the lung by IP injection and no phage was detected after 12 h. The importance of phage dose on the pharmacokinetics/pharmacodynamics (PK/PD) of inhaled phage therapy was recently confirmed by Chow et al. using *Pseudomonas* phage PEV31 [30].

Carrigy et al. recently demonstrated the prophylactic function of nebulized D29 phage for protection against *Mycobacterium tuberculosis* infection in a mouse model [31]. Phage was delivered with a vibrating mesh nebulizer and a dose of 6.6 log phage reached the lung and remained there for 90 min post-delivery, suggesting that phage was not rapidly cleared in the mouse lung. Low doses of *M. tuberculosis* (5–100 CFU) were given to mice 30 min post phage administration. This phage pretreatment was able to significantly reduce the bacterial burden in mouse lungs at 24 h and 3 weeks post infection. The prophylactic effect of phage was also demonstrated in a rat model against methicillin-resistant *S. aureus* infection [32]. Phage was given by a vibrating mesh nebulizer 4 h before the bacterial challenge, higher survival rate (60–70% improvement) with a 2 log bacterial reduction in the rat lungs were observed. Both studies demonstrated prophylactic treatment with sufficient dose of nebulized phage may provide protection to immunocompromised individuals and health care professionals who are at risk of exposure to "superbugs".

There is accumulating evidence that bacterial clearance by phage therapy requires the synergy between phage and host immune system. Therefore, the translation of preclinical data collected from rodent to humans should be treated with care due to the significant difference in their immune systems [33]. Cao et al. explored the phage antibacterial effect of hemorrhagic pneumonia in a mink model [27]. Effective treatment outcomes were achieved at multiplicity of infection (MOI) of 10 with an 80% survival rate at 12 days after phage administrated by means of ultrasonic nebulization.

4. Dry powder inhalers

4.1 Powder formulation

Although nebulization has been the method of choice for phage delivery in treating lung infections in clinical settings, dry powder formulations are preferred to liquid formulations in terms of storage, transportation and administration [34]. Compared to nebulizers, dry powder inhalers (DPIs) are easier to handle without the need of a power source, fewer cleaning requirements and quick delivery [35]. Current research on pharmaceutical development of inhaled phage dry powder mainly focuses on formulation optimization for sufficient powder dispersibility to deliver phage to the lung and storage stability. The choice of excipients plays a key role among all the techniques to produce phage dry powder. Zhang et al. published a comprehensive review to discuss how the choice of excipients affecting the stability of phage in the solid-state [36]. Overall, sucrose, lactose and trehalose are the most popular disaccharides in phage powder formulations. Freeze drying (FD), spray drying (SD) and spray freeze drying (SFD) have been used to generate inhalable phage dry powders with these excipients.

FD is a commonly employed technique to stabilize drugs in solid state [37]. Puapermpoonsiri et al. used FD to generate dry powder of phage-loaded poly(lactic-co-glycolic acid) (PLGA) microspheres designed for pulmonary delivery [38]. Although phages were successfully incorporated into the PLGA microparticles, the poor shelf-life of the encapsulated phage which completely deactivated within 7 days either stored at 4 °C or 22 °C was discouraging. In their follow-up study, they investigated the feasibility of using a high concentration of sucrose (0.5 M) or PEG6000 (5%) to stabilize the FD phage cake [39]. Although rapid phage reduction was still noted over the first 7–14 days, phage remained relatively stable in the powder formulations thereafter. Since then, a number of studies have studied the impacts of various excipients on the production loss and storage stability of FD phages [40–43]. Among all excipients examined, sucrose and trehalose were identified as the most promising stabilizers to preserve phage viability upon the dehydration in the drying process and upon storage. The residual moisture content was found to play an important role in maintaining phage stability. Similar to other protein therapeutics, a 3–6% moisture content of the powder cake was found to be optimal for phage preservation [39, 41]. Although the mechanisms of phage stabilization in dry powder by these sugars are still unclear. Two most acceptable hypotheses for the stabilization of proteins in the solid state by sugars are water replacement and vitrification, which may also be applicable to phages because they are mostly composed of proteins.

In general, FD powder is not respirable, and a separate milling step is required to reduce the particle size to <5 μm, suitable for pulmonary delivery. However, the high-energy milling may cause additional phage loss due to the generation of heat and mechanical stresses. Golshahi et al. prepared FD formulations of KS4-M and ΦKZ phages with 60% lactose and 40% lactoferrin suitable for pulmonary delivery without milling [44]. The size of the phage powder was within the inhalable range (< 5 μm) and acceptable aerosol performance with a fine particle dose of $>10^6$ pfu using an Aerolizer was achieved. The production loss was 1–2 log which was not desirable, but the FD phage powders were stable with negligible titer reduction within 3 months storing either at 4 °C or 22 °C.

SD is a well-established single-step technique employed for the production of many inhaled pharmaceutical products [45]. Matinkhoo et al. were among the first to study the feasibility of using SD to produce inhalable phage powders comprising trehalose and leucine with or without a third excipients (a surfactant or casein sodium salt) [46]. In these formulations, trehalose was used to protect phage against dehydration; leucine forming a crystalline shell at the particle surface was used to enhance the dispersibility of powders; and a surfactant was employed to reduce aggregation of phage during the drying process. Due to the thermal sensitivity of phage, a low drying temperature was used to produce SD powders with acceptable production loss (0.4–0.8 log) and phage lung dose (7–8 log pfu). Trehalose-alone formulation was employed by Vandenheuvel et al., but the production loss was found to be phage dependent [47]. On the other hand, trehalose-leucine and lactose-leucine systems could stabilize a panel of *Pseudomonas* phage upon the SD process [48–51]. Since the SD trehalose and lactose is amorphous, Chang et al. demonstrated that the addition of a sufficient amount of leucine (at least 20%) was critical to stabilize phage by minimizing recrystallization of trehalose/lactose during powder production process [48]. Despite a low production loss was achieved, particle merging was still significant for formulation containing 80% sugar and 20% leucine due to moisture sorption upon handling. Therefore, higher leucine content and the addition of mannitol to the excipient system was attempted to improve the morphology and reduce the moisture sorption capacity of the phage powders during handling and storage (**Figure 1a-c**) [49, 50]. Although these approaches significantly reduce the problem of particle merging and make powder handling easier, they failed to stop the recrystallization of the amorphous content at high humidity conditions (RH > 50%). Therefore, storing the SD powders at low humidity conditions (RH ≤ 20%) was generally recommended [48, 49, 52, 53]. The storage temperature was also reported to be important on phage dry powder stability. It is generally recommended to store phage drug powder at a temperature at least 50 °C below the glass transition temperature (Tg) of the powders [54].

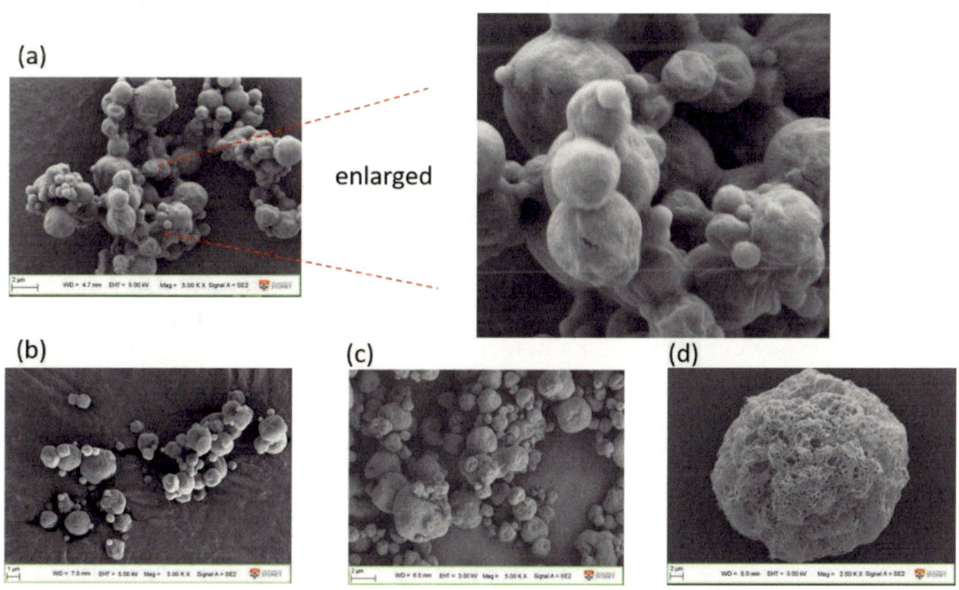

Figure 1.
Representative scanning electron microscopy images of phage powders produced by spray drying (a-c) and spray freeze drying (d). (a) 80% trehalose+20% leucine; (b) 60% trehalose +20% mannitol +20% leucine; (c) 70% trehalose +30% leucine and (d) 60% trehalose+20% mannitol and 20% leucine.

Overall, SD phage powders composed of trehalose/lactose not less than 40% of the total solid content together with leucine and mannitol was able to stabilize phage in powder form with sufficient long shelf-life (≤ 1 log titer loss in 12 months) under refrigeration or room temperature at RH < 20% and yield acceptable lung dose (10^5–10^7 pfu) [46–50, 53]. While leucine is a commonly employed surface active agent to improve the powder dispersity of inhaled pharmaceuticals, trileucine has also been increasingly used to improve aerosol performance and stability of SD powders for inhalation. Recently, Carrigy et al. demonstrated the effectiveness of a trileucine and trehalose system in preserving an anti-Campylobacter phage, CP30A, in powder form for long-distance ambient temperature transportation [55, 56].

SFD is a relatively new drying technique to produce inhalable dry powders. The produced powders are superior to those prepared by traditional FD in terms of structure, quality, and the retention of volatiles and bioactive compounds [57]. The suitability of SFD porous mannitol carriers for pulmonary delivery of drug nanoparticles and biologics have been demonstrated [58–60]. Leung et al. produced SFD phage powder and compared their differences of powder properties with the SD phage powders (**Figure 1d**). With the use of a high frequency of ultrasonic nozzle in the SFD process, a significant titer reduction (>2 log) was noted in the spraying process, making the overall production loss inferior compared with the SD process [53]. Nonetheless, the larger porous carrier provided a larger extent of protection of the embedded phage during aerosolization with a higher recovery of viable phage compared with the SD counterparts. The conventional SFD process is a two-step manufacturing process, which hinders scaling up. Ly et al. used an atmospheric spray freeze-drying (ASFD) technique, which is a single step process, to prepare D29 phage powder [61]. An acceptable titer loss (~0.6 log) was noted due to the use of a twin-fluid nozzle and improved mass and heat transfer rates.

4.2 *In vivo* efficacy of inhalable phage dry powder

Pulmonary delivery of dry powder to small animals is challenging as they cannot inhale powder actively. Intratracheal delivery using a dry powder insufflator, either the commercially available Penn-Century models or custom-made insufflators [62], are commonly employed to introduce powders directly into the lungs of the experimental animals. Chang et al. explored the *in vivo* efficacy of phage powder to treat lung infections caused by MDR *P. aeruginosa* in a mice pneumonia model [63]. After challenging the neutropenic mice with intratracheal administration of the bacterial suspension for 2 h, powder of phage PEV20 was administrated use a Penn-Century dry powder insufflator at a concentration of 2×10^7 pfu/mg. A significant bacterial reduction (5.3 log cfu) was noted after 24 h post-infection accompanies with 1 log phage propagation. The successful treatment outcomes and safety profile from this study warrant further investigation to fully evaluate the therapeutic potential of inhaled phage powder in managing lung infections.

5. Other inhalation devices

5.1 Metered dose inhaler

Pressurized metered-dose inhalers (pMDIs) are the most popular inhalers for the treatment of asthma and chronic obstructive pulmonary diseases. To date, only one study has attempted this type of device to aerosolize phage [64]. The phage cocktail suspension containing FKZ/D3 and KS4-M phages, was formulated in a reverse emulsion with Tyloxapol surfactant using hydrofluoroalkane 134a as the propellant. A limited loss of phage activity (0.5–0.9 log) upon the actuation was observed, but the long term storage stability of the phages was not assessed. Further studies to examine the interactions between phage and liquefied propellant gas [65], and maximum loading capacity of phage/puff are required to move this inhaler choice forward.

5.2 Soft mist inhaler

Soft mist inhaler (SMI) is a relatively new generation, propellant-free inhaler that delivers drugs to the lung more efficiently than pMDIs because of the lower spray velocity and longer duration time [66]. Carrigy et al. compared the delivery efficiency of phage among vibrating mesh nebulizer, jet nebulizer and SMI [14]. SMI was showed to deliver phage D29 at high titers quickly (~5×10^8 pfu/actuation) with an acceptable titer reduction (0.6 log pfu/ml) and a higher lung delivery (3.2×10^6 pfu/actuation of inhalable active phage). This compact and light weight device may act as an attractive option for self-administration of phage aerosols.

6. Combination of phage therapy and antibiotic to treat lung infections

6.1 Mechanisms of phage-antibiotic synergy

With the emergence of phage-resistant bacteria [67], the combination therapy of antibiotics and phages has drawn increasing attention. Synergistic effect of antibiotic and phage against *S. aureus* was first reported by Himmelweit et al. back in 1945 [68]. Similar synergistic antibacterial effects have also been observed in a number of subsequent studies [69–79]. In 2007, Comeau et al., coined the term phage-antibiotic synergy (PAS) corresponding to an incident where the killing effect of bacterial strains considerably higher when phage production increases by the sublethal concentrations of particular antibiotics [80]. While many antibiotics exhibit synergistic effect in combination with phages, two specific classes of antibiotics (namely beta-lactams and fluoro-quinolones) were shown to produce a more consistent and pronounced antibacterial synergistic effect with phage therapy. The precise mechanisms contributing to phage-antibiotic synergy remain largely unknown. A few possible mechanisms have been proposed (**Figure 2**): (1) Antibiotic causes cell elongation or filamentation, thus subsequently promoting phage production; (2) Degradation of the extracellular membrane of bacteria by phage facilitates internalization of antibiotic into cells; (3) Auto-aggregation of bacterial cells leads to synergism; (4) Bacteria containing complete prophages could be induced by antibiotics which further kill bacteria [81]. The capacity of phage in resensitizating bacteria to certain antibiotics have also been reported as the host bacteria cannot develop resistance to phage and antibiotic simultaneously [82–84]. As a result, the phage-antibiotic combination can kill both phage-sensitive and antibiotic-sensitive pathogens with the phage lysing cells resistant to antibiotics and antibiotic mediated killing of phage-resistant bacterial cells and eventually inhibit the infections.

Interestingly, the sequence of phage and antibiotic administration was found to be critical in the overall antibacterial effect from the combination treatment. Chaudhry et al. showed the efficiency of removing *P. aeruginosa* PA14 biofilm was higher when the biofilm was treated with phages before antibiotics [85]. A similar observation was also reported in another study evaluating phage-antibiotics combination therapy against *S. aureus* biofilms [86]. However, the observed synergistic effects were found to be dependent on the class of antibiotics used. Pre-treatment

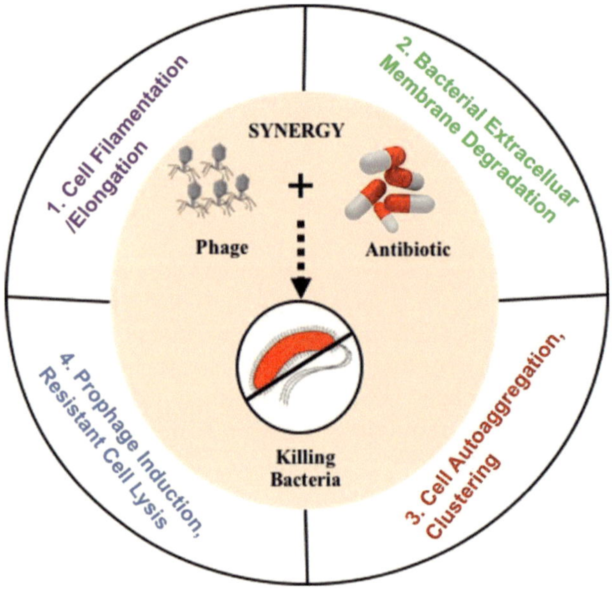

Figure 2.
Possible mechanisms responsible for phage-antibiotic synergy.

with phage led to favorable antibacterial effect when combined with linezolid or tetracycline, whereas antagonism was observed between the phage and dicloxacillin or cefazolin. Furthermore, it is noteworthy that an antagonistic effect was observed when the bacterial biofilm was treated with antibiotics preceding the phage therapy, irrespective of which class of antibiotics used [86].

6.2 Novel tools for selection of optimum phage-antibiotic combination

Since the exact mechanisms responsible for PAS are still unclear and the choice of the combinations is mostly empirical, it is not surprising that mixed results were reported in the literature [72, 82]. Also, the concentration of antibiotics used in previous studies was limited to one or two levels, which is not enough to predict the efficacious concentration when applied in clinical treatment. To solve these problems, Liu et al. developed a high-throughput platform called synogram by combining an optically based real-time microtiter plate readout with a matrix-like heat map to quickly assess the effects of various phage and antibiotic concentrations on bacterial growth [87]. They concluded that PAS is highly dependent on the antibacterial mechanism of action for antibiotic and phage pairs and their stoichiometry.

To guide the choice of phage-antibiotic combination, Rodriguez-Gonzalez et al. [88] developed an *in silico* nonlinear population dynamics model taking into account the systemic interactions between bacteria, phage and antibiotics to mimic *in vivo* application by given an immune response against bacteria. Using two *P. aeruginosa* strains, one phage-sensitive (resistant to antibiotic) and one antibiotic sensitive (resistant to phage), as the model bacteria, the phage-antibiotic combination therapy was confirmed to outperform the monotherapy. The role of the host immune response was also evaluated and the model predicted that the phage-antibiotic combination failed to eliminate the infection when innate immunity was removed or severely reduced. Their findings confirmed the clearance of infection is depending on the nonlinear synergistic interactions between phage, antibiotic, and innate immunity. The *in silico* prediction was consistent with previous experimental results obtained *in vitro* and *in vivo*. While this model is a valuable tool in identifying potential phage-antibiotic combinations, further modification of the model to yield high-resolution temporal data in addition to the final results will be useful for quantitative comparison of the model-based predictions with experimental results.

6.3 Formulations of phage-antibiotic combination to treat lung infections

Streptococcus pneumoniae, S. aureus, B. cepacia complex, Klebsiella pneumonia and *P. aeruginosa* are the major causative pathogens for lung infections. A summary on previous work on the combination phage-antibiotic therapy against these pathogens were provided in Chang et al. [8]. Recently, Lin et al. screened a panel of antibiotics with PEV20 phage to target two *P. aeruginosa* strains and ciprofloxacin showed the highest synergistic effects. The combination was then nebulized using a jet nebulizer and a mesh nebulizer with no difference in the antibacterial effect observed between the nebulized samples and non-nebulized suspension [89]. Later, the same research team investigated the feasibility of formulating this combination into dry powder formulations [90]. PEV20 phage and ciprofloxacin were co-spray dried with leucine and with or without lactose. Both formulations maintained bactericidal synergy after dispersion using a low resistance inhaler or a high resistance inhaler, both showing acceptable FPF (60–75%). The antimicrobial efficacy of the PEV20-ciprofloxacin combination powder was also confirmed in a mice respiratory infection model with significant bacteria reduction (5.9 log) at 24 h post-treatment, while no loss of bacteria viability when mice was treated with phage or antibiotics alone [91]. The long-term storage stability of the combination powder at 4 °C and 20% R.H. was also confirmed [92].

7. Challenges for pulmonary delivery of phage and future perspective

Phage therapy is evolving as a promising alternative or an adjuvant to antibiotics for the battle against MDR bacteria. Although a few randomized, double-blind and placebo-controlled clinical trials have been conducted to assess tolerance and/or efficacy of phage therapy in the past few years, none of the completed trials have yielded data supporting the promising observations noted in the experimental phage therapy conducted in animals and humans. Górski et al. highlighted the importance of the quality and titer of the phage preparations and their delivery efficiency to the target sites to ensure a sufficient high phage to bacteria concentration in the vicinity of infected tissues [93]. For lung infection, directly delivering phage preparation to the airways enhance the incidence of phage getting access to its host bacteria, avoiding the rapid clearance in systemic circulation. Advancements have been made in the past decade to improve the formulations for pulmonary delivery of phage. Here we highlight some hurdles remained to be tackled to bring inhaled phage therapy to clinical settings beyond compassionate use and a few prospective research directions for the commercial application of aerosol formulations.

As a sufficient amount of phage at the site of infection is the prerequisite for successful therapy, nebulizers and DPI are better choice for pulmonary delivery of phage compared with pMDI and SMI due to their capacity of high dose delivery. The detrimental effect of the various type of nebulizers to phage was found to be phage-specific, likely attributing to the tail morphology of phage [21] and compositions of the phage formulations [18]. Systematic studies to confirm their impacts on phage nebulization will provide important information in developing new phage cocktail formulations. Although liquid formulations are commonly used for phage therapy, solid phage formulations are more desirable for long-term storage and transportation.

While stable phage powder formulations have been successfully achieved with storage at ambient temperature, they are usually required to be handled and stored at low humidity conditions (RH < 20%) [48–50]. These would be easily achievable in a manufacturing setting and with pharmaceutical packaging designs. As excessive environmental moisture could also be relevant in patients' homes or in healthcare settings, the impacts of humidity on powdered phage administration should be evaluated to ensure the phage product could be used successfully in different geographic regions over the world. In preparing phage-powder formulation, trehalose, lactose, and leucine are commonly employed to stabilize phage. However, these excipients have not been approved for inhalation except lactose was approved as a carrier which is not expected to be delivered to the lower respiratory tract. Further *in vivo* studies are required to evaluate the safety profile of these excipients for both short term and long term usage.

Currently, *in vivo* data of phage therapy for lung infections mostly focused on acute infections that phage preparation was given at within a few hours post-infection. However, in clinical settings, the phages are unlikely given immediately after the onset of infection, the postponed treatment may lead to significant bacterial growth and biofilm formation, more research is needed to evaluate the therapeutic efficacy of phage therapy against chronic lung infection in animal models. Moreover, more extensive *in vivo* PKPD evaluations are needed to investigate the optimal administration dose and time for pulmonary phage therapy.

The role of the immune system on phage therapy is largely unexplored in animal studies and human trials [33, 88]. Depending on administration route, phage type and phage dose, and duration of phage therapy can lead to the generation of neutralizing antibodies [94]. Together with increasing evidences showing the interactions between phage and mammalian cells [95–97], it would be worthwhile to explore the interaction between phage formulations with lung leukocytes and epithelial cells lining the alveolar surface and the conducting airways.

Current phage formulation research is largely empirical based. To speed up the research progress for phage therapy, *in silico* models and database would be required to predict phage-excipient interaction, phage-antibiotic combination and pharmacokinetic/pharmacodynamics (PKPD) profiles.

8. Conclusion

In the past decade, highly acceptable formulations have been achieved with minimal phage loss and desirable stability for pulmonary delivery using both nebulizers and dry powder inhalers. The synergistic effect of the phage-antibiotic combination provides an efficient way to prevent the emergence of bacterial resistance and reduce the toxicity of antibiotic use. However, systematic PKPD profile of phage after administration by inhalation, and the modern tools to accurately predict the result of combination therapy are still pending. With the advent of phage research, the sound manufacturing and regulatory guidelines towards successful clinical trials to bring phage therapy to clinical settings will be beneficial to the patients suffering from bacterial infections.

Acknowledgements

The authors gratefully acknowledge the provision of graduate studentship from CUHK to W. Yan and S. Mukhopadhyay is supported by the HKPFS. The funding support from University Grants Committee Hong Kong (ref. 24300619) for our phage research is greatly acknowledged.

Author details

Wei Yan, Subhankar Mukhopadhyay, Kenneth Kin Wah To
and Sharon Shui Yee Leung*
Faculty of Medicine, School of Pharmacy, The Chinese University of Hong Kong,
Shatin, Hong Kong SAR, China

*Address all correspondence to: sharon.leung@cuhk.edu.hk

References

[1] Forum of International Respiratory Societies, European Respiratory Society. The global impact of respiratory disease. 2017.

[2] Cookson WOCM, Cox MJ, Moffatt MF. New opportunities for managing acute and chronic lung infections. Nat Rev Microbiol 2018;16:111-120. DOI: 10.1038/nrmicro.2017.122

[3] Qureshi ZA, Hittle LE, O'Hara JA, Rivera JI, Syed A, Shields RK, Pasculle AW, Ernst RK, Doi Y. Colistin-resistant Acinetobacter baumannii: beyond carbapenem resistance. Clin Infect Dis 2015;60:1295-1303. DOI: 10.1093/cid/civ048

[4] Bialvaei AZ, Kafil HS. Colistin, mechanisms and prevalence of resistance. Curr Med Res Opin 2015;31:707-721. DOI: 10.1185/03007995.2015.1018989

[5] Ah Y-M, Kim A-J, Lee J-Y. Colistin resistance in Klebsiella pneumoniae. Int J Antimicrob Agents. 2014;44:8-15. DOI: 10.1016/j.ijantimicag.2014.02.016

[6] Abedon ST. Phage therapy of pulmonary infections. Bacteriophage 2015;5:e1020260. https://doi.org/10.1080/21597081.2015.1020260.

[7] Melo LDR, Oliveira H, Pires DP, Dabrowska K, Azeredo J. Phage therapy efficacy: a review of the last 10 years of preclinical studies. Crit Rev Microbiol. 2020;46:78-99. DOI: 10.1080/1040841X.2020.1729695

[8] Chang RYK, Wallin M, Lin Y, Leung SSY, Wang H, Morales S, Chan HK. Phage therapy for respiratory infections. Adv Drug Deliv Rev. 2018;133:76-86. DOI: 10.1016/j.addr.2018.08.001.

[9] Jault P, Leclerc T, Jennes S, Pirnay JP, Que YA, Resch G, Rousseau AF, Ravat F, Carsin H, Le Floch R, Schaal JV, Soler C, Fevre C, Arnaud I, Bretaudeau L, Gabard J. Efficacy and tolerability of a cocktail of bacteriophages to treat burn wounds infected by Pseudomonas aeruginosa (PhagoBurn): a randomised, controlled, double-blind phase 1/2 trial. Lancet Infect Dis. 2019;19:35-45. DOI: 10.1016/S1473-3099(18)30482-1

[10] Hoyle N, Zhvaniya P, Balarjishvili N, Bolkvadze D, NadareishviliL,NizharadzeD,WittmannJ, Rohde C, Kutateladze M. Phage therapy against Achromobacter xylosoxidans lung infection in a patient with cystic fibrosis: a case report. Res Microbiol. 2018;169:540-542. DOI: 10.1016/j.resmic.2018.05.001

[11] Lebeaux D, Merabishvili M, Caudron E, Lannoy D, Van Simaey L, DuyvejonckH,GuillemainR,ThumerelleC, Podglajen I, Compain F, Kassis N, MainardiJL,WittmannJ,RohdeC,PirnayJP, Dufour N, Vermeulen S, Gansemans Y, Nieuwerburgh FV, Vaneechoutte M. A case of phage therapy against pandrug-resistant Achromobacter xylosoxidans in a 12-year-old lung-transplanted cystic fibrosis patient. Viruses. 2021;13:60. DOI: 10.3390/v13010060

[12] Kutateladze M, Adamia R. Phage therapy experience at the Eliava Institute. Med Mal Infect. 2008;38:426-430. DOI: 10.1016/j.medmal.2008.06.023

[13] Aslam S, Courtwright AM, Koval C, Lehman SM, Morales S, Furr C-LL, Rosas F, Brownstein MJ, Fackler JR, Sisson BM, Biswas B, Henry M, Luu T, Bivens BN, Hamilton T, Duplessis C, Logan C, Law N, Yung G, Turowski J, Anesi J, Strathdee SA, Schooley RT. Early clinical experience of bacteriophage therapy in 3 lung transplant recipients. Am J Transplant. 2019;19:2631-2639. DOI: 10.1111/ajt.15503

[14] Cooper CJ, Denyer SP, Maillard J-Y. Stability and purity of a bacteriophage cocktail preparation for nebulizer delivery. Lett Appl Microbiol. 2014;58:118-122. DOI: 10.1111/lam.12161

[15] Carrigy NB, Chang RY, Leung SSY, Harrison M, Petrova Z, Pope WH, Hatfull GF, Britton WJ, Chan HK, Sauvageau D, Finlay WH, Vehring R. Anti-tuberculosis bacteriophage D29 delivery with a vibrating mesh nebulizer, jet nebulizer, and soft mist inhaler. Pharm Res. 2017;34:2084-2096. DOI: 10.1007/s11095-017-2213-4

[16] Golshahi L, Seed KD, Dennis JJ, Finlay WH. Toward modern inhalational bacteriophage therapy: nebulization of bacteriophages of Burkholderia cepacia complex. J Aerosol Med Pulm Drug Deliv. 2008;21:351-360. DOI: 10.1089/jamp.2008.0701

[17] Bodier-Montagutelli E, Morello E, L'Hostis G, Guillon A, Dalloneau E, Respaud R, Pallaoro N, Blois H, Vecellio L, Gabard J, Heuzé-Vourc'h N. Inhaled phage therapy: a promising and challenging approach to treat bacterial respiratory infections. Expert Opin Drug Deliv. 2017;14:959-972. DOI: 10.1080/17425247.2017.1252329

[18] Semler DD, Goudie AD, Finlay WH, Dennis JJ. Aerosol phage therapy efficacy in Burkholderia cepacia complex respiratory infections. Antimicrob Agents Chemother. 2014;58:4005-4013. DOI: 10.1128/AAC.02388-13

[19] Liu K, Wen Z, Li N, Yang W, Wang J, Hu L, Dong X, Lu J, Li J. Impact of relative humidity and collection media on mycobacteriophage D29 aerosol. Appl Environ Microbiol. 2012;78:1466-1472. DOI: 10.1128/AEM.06610-11

[20] Liu K, Yang W, Dong X, Cong L, Li N, Li Y, Wen Z, Yin Z, Lan Z, Li W, Li J. Inhalation study of Mycobacteriophage D29 aerosol for mice by endotracheal route and nose-only exposure. J Aerosol Med Pulm Drug Deliv. 2016;29:393-405. DOI: 10.1089/jamp.2015.1233

[21] Verreault D, Marcoux-Voiselle M, Turgeon N, Moineau S, Duchaine C. Resistance of aerosolized bacterial viruses to relative humidity and temperature. Appl Environ Microbiol. 2015;81:7305-7311. DOI: 10.1128/AEM.02484-15

[22] Leung SSY, Carrigy NB, Vehring R, Finlay WH, Morales S, Carter EA, Britton WJ, Kutter E, Chan HK. Jet nebulization of bacteriophages with different tail morphologies – Structural effects. Int J Pharm. 2019;554:322-326. DOI: 10.1016/j.ijpharm.2018.11.026

[23] Sahota JS, Smith CM, Radhakrishnan P, Winstanley C, Goderdzishvili M, Chanishvili N, Kadioglu A, Callaghan C, Clokie MRJ. Bacteriophage delivery by nebulization and efficacy against phenotypically diverse Pseudomonas aeruginosa from cystic fibrosis patients. J Aerosol Med Pulm Drug Deliv. 2015;28:353-360. DOI: 10.1089/jamp.2014.1172

[24] Turgeon N, Toulouse M-J, Martel B, Moineau S, Duchaine C. Comparison of five bacteriophages as models for viral aerosol studies. Appl Environ Microbiol. 2014;80:4242-4250. DOI: 10.1128/AEM.00767-14

[25] Astudillo A, Leung SSY, Kutter E, Morales S, Chan H-K. Nebulization effects on structural stability of bacteriophage PEV 44. Eur J Pharm Biopharm. 2018;125:124-130. DOI: 10.1016/j.ejpb.2018.01.010

[26] Pritchard JN, Hatley RH, Denyer J, Hollen D von. Mesh nebulizers have become the first choice for new nebulized pharmaceutical drug developments. Ther Deliv.

2018;9:121-136. DOI: 10.4155/tde-2017-0102

[27] Cao Z, Zhang J, Niu YD, Cui N, Ma Y, Cao F, Jin L, Li Z, Xu Y. Isolation and characterization of a "phiKMV-like" bacteriophage and its therapeutic effect on mink hemorrhagic pneumonia. Plos One. 2015;10:e0116571. DOI: 10.1371/journal.pone.0116571

[28] Marqus S, Lee L, Istivan T, Kyung Chang RY, Dekiwadia C, Chan H-K, Yeo LY. High frequency acoustic nebulization for pulmonary delivery of antibiotic alternatives against Staphylococcus aureus. Eur J Pharm Biopharm. 2020;151:181-188. DOI: 10.1016/j.ejpb.2020.04.003

[29] Carmody LA, Gill JJ, Summer EJ, Sajjan US, Gonzalez CF, Young RF, LiPuma JJ. Efficacy of bacteriophage therapy in a model of Burkholderia cenocepacia pulmonary infection. J Infect Dis. 2010;201:264-271. DOI: 10.1086/649227

[30] Chow MYT, Chang RYK, Li M, Wang Y, Lin Y, Morales S, McLachlan AJ, Kutter E, Li J, Chan HK. Pharmacokinetics and time-kill study of inhaled antipseudomonal bacteriophage therapy in mice. Antimicrob Agents Chemother. 2020;65. DOI: 10.1128/AAC.01470-20

[31] Carrigy NB, Larsen SE, Reese V, Pecor T, Harrison M, Kuehl PJ, Hatfull GF, Sauvageau D, Baldwin SL, Finlay WH, Coler RN, Vehring R. Prophylaxis of Mycobacterium tuberculosis H37Rv infection in a preclinical mouse model via inhalation of nebulized bacteriophage D29. Antimicrob Agents Chemother. 2019;63. DOI: 10.1128/AAC.00871-19

[32] Prazak J, Valente L, Iten M, Grandgirard D, Leib SL, Jakob SM, Haenggi M, Que YA, Cameron DR. Nebulized bacteriophages for prophylaxis of experimental ventilator-associated pneumonia due to methicillin-resistant Staphylococcus aureus. Crit Care Med. 2020;48:1042-1046. DOI: 10.1097/CCM.0000000000004352

[33] Roach DR, Leung CY, Henry M, Morello E, Singh D, Di Santo JP, Weitz JS, Debarbieux L. Synergy between the host immune system and bacteriophage is essential for successful phage therapy against an acute respiratory pathogen. Cell Host Microbe 2017;22:38-47.e4. DOI: 10.1016/j.chom.2017.06.018.

[34] Emerging inhalation aerosol devices and strategies: Where are we headed? Adv Drug Deliv Rev. 2014;75:3-17. DOI: 10.1016/j.addr.2014.03.006

[35] Respaud R, Vecellio L, Diot P, Heuzé-Vourc'h N. Nebulization as a delivery method for mAbs in respiratory diseases. Expert Opin Drug Deliv. 2015;12:1027-1039. DOI: 10.1517/17425247.2015.999039

[36] Zhang Y, Zhang H, Ghosh D. The stabilizing excipients in dry state therapeutic phage formulations. AAPS PharmSciTech. 2020;21:133. DOI: 10.1208/s12249-020-01673-5

[37] Schwegman JJ, Hardwick LM, Akers MJ. Practical formulation and process development of freeze-dried products. Pharm Dev Technol 2005;10:151-173. DOI: 10.1081/PDT-56308.

[38] Puapermpoonsiri U, Spencer J, van der Walle CF. A freeze-dried formulation of bacteriophage encapsulated in biodegradable microspheres. Eur J Pharm Biopharm. 2009;72:26-33. DOI: 10.1016/j.ejpb.2008.12.001

[39] Puapermpoonsiri U, Ford SJ, van der Walle CF. Stabilization of bacteriophage during freeze drying. Int J Pharm. 2010;389:168-175. DOI: 10.1016/j.ijpharm.2010.01.034

[40] Alfadhel M, Puapermpoonsiri U, Ford SJ, McInnes FJ, van der Walle CF. Lyophilized inserts for nasal administration harboring bacteriophage selective for Staphylococcus aureus: In vitro evaluation. Int J Pharm. 2011;416:280-287. DOI: 10.1016/j.ijpharm.2011.07.006

[41] Zhang Y, Peng X, Zhang H, Watts AB, Ghosh D. Manufacturing and ambient stability of shelf freeze dried bacteriophage powder formulations. Int J Pharm. 2018;542:1-7. DOI: 10.1016/j.ijpharm.2018.02.023

[42] Dini C, de Urraza PJ. Effect of buffer systems and disaccharides concentration on Podoviridae coliphage stability during freeze drying and storage. Cryobiology. 2013;66:339-342. DOI: 10.1016/j.cryobiol.2013.03.007

[43] Merabishvili M, Vervaet C, Pirnay J-P, Vos DD, Verbeken G, Mast J, Chanishvili N, Vaneechoutte M. Stability of Staphylococcus aureus phage ISP after freeze-drying (lyophilization). PLos One. 2013;8. DOI: 10.1371/journal.pone.0068797

[44] Golshahi L, Lynch KH, Dennis JJ, Finlay WH. In vitro lung delivery of bacteriophages KS4-M and ΦKZ using dry powder inhalers for treatment of Burkholderia cepacia complex and Pseudomonas aeruginosa infections in cystic fibrosis. J Appl Microbiol. 2011;110:106-117. DOI: 10.1111/j.1365-2672.2010.04863.x

[45] Vehring R. Pharmaceutical particle engineering via spray drying. Pharm Res. 2008;25:999-1022. DOI: 10.1007/s11095-007-9475-1

[46] Matinkhoo S, Lynch KH, Dennis JJ, Finlay WH, Vehring R. Spray-dried respirable powders containing bacteriophages for the treatment of pulmonary infections. J Pharm Sci. 2011;100:5197-5205. DOI: 10.1002/jps.22715

[47] Vandenheuvel D, Singh A, Vandersteegen K, Klumpp J, Lavigne R, Van den Mooter G. Feasibility of spray drying bacteriophages into respirable powders to combat pulmonary bacterial infections. Eur J Pharm Biopharm. 2013;84:578-582. DOI: 10.1016/j.ejpb.2012.12.022

[48] Chang RY, Wong J, Mathai A, Morales S, Kutter E, Britton W, Li J, Chan HK. Production of highly stable spray dried phage formulations for treatment of Pseudomonas aeruginosa lung infection. Eur J Pharm Biopharm. 2017;121:1-13. DOI: 10.1016/j.ejpb.2017.09.002

[49] Leung SSY, Parumasivam T, Gao FG, Carter EA, Carrigy NB, Vehring R, Finlay WH, Morales S, Britton WJ, Kutter E, Chan HK. Effects of storage conditions on the stability of spray dried, inhalable bacteriophage powders. Int J Pharm. 2017;521:141-149. DOI: 10.1016/j.ijpharm.2017.01.060

[50] Leung SSY, Parumasivam T, Nguyen A, Gengenbach T, Carter EA, Carrigy NB, Wang H, Vehring R, Finlay WH, Morales S, Britton WJ, Kutter E, Chan HK. Effect of storage temperature on the stability of spray dried bacteriophage powders. Eur J Pharm Biopharm. 2018;127:213-222. DOI: 10.1016/j.ejpb.2018.02.033

[51] Chang RYK, Wallin M, Kutter E, Morales S, Britton W, Li J, Chan HK. Storage stability of inhalable phage powders containing lactose at ambient conditions. Int J Pharm. 2019;560:11-18. DOI: 10.1016/j.ijpharm.2019.01.050

[52] Vandenheuvel D, Meeus J, Lavigne R, Van den Mooter G. Instability of bacteriophages in spray-dried trehalose powders is caused by crystallization of the matrix. Int J Pharm. 2014;472:202-205. DOI: 10.1016/j.ijpharm.2014.06.026

[53] Leung SSY, Parumasivam T, Gao FG, Carrigy NB, Vehring R, Finlay WH,

Morales S, Britton WJ, Kutter E, Chan HK. Production of inhalation phage powders using spray freeze drying and spray drying techniques for treatment of respiratory infections. Pharm Res. 2016;33:1486-1496. DOI: 10.1007/s11095-016-1892-6

[54] Chang RYK, Kwok PCL, Khanal D, Morales S, Kutter E, Li J, Chan HK. Inhalable bacteriophage powders: Glass transition temperature and bioactivity stabilization. Bioeng Transl Med. 2020;5:e10159. DOI: 10.1002/btm2.10159

[55] Carrigy NB, Liang L, Wang H, Kariuki S, Nagel TE, Connerton IF, Vehring R. Trileucine and pullulan improve anti-Campylobacter bacteriophage stability in engineered spray-dried microparticles. Ann Biomed Eng. 2020;48:1169-1180. DOI: 10.1007/s10439-019-02435-6

[56] Carrigy NB, Liang L, Wang H, Kariuki S, Nagel TE, Connerton IF, et al. Spray-dried anti-Campylobacter bacteriophage CP30A powder suitable for global distribution without cold chain infrastructure. Int J Pharm. 2019;569:118601. DOI: 10.1016/j.ijpharm.2019.118601

[57] Ishwarya SP, Anandha ramakrishnan C, Stapley AGF. Spray-freeze-drying: A novel process for the drying of foods and bioproducts. Trends Food Sci Technol. 2015;41:161-181. DOI: 10.1016/j.tifs.2014.10.008

[58] Leung SSY, Wong J, Guerra HV, Samnick K, Prud'homme RK, Chan H-K. Porous mannitol carrier for pulmonary delivery of cyclosporine A nanoparticles. AAPS J. 2017;19:578-586. DOI: 10.1208/s12248-016-0039-3

[59] Okuda T, Morishita M, Mizutani K, Shibayama A, Okazaki M, Okamoto H. Development of spray-freeze-dried siRNA/PEI powder for inhalation with high aerosol performance and strong pulmonary gene silencing activity. J Control Release Off J Control Release Soc. 2018;279:99-113. DOI: 10.1016/j.jconrel.2018.04.003

[60] Fukushige K, Tagami T, Naito M, Goto E, Hirai S, Hatayama N, Yokota H, Yasui T, Baba Y, Ozeki T. Developing spray-freeze-dried particles containing a hyaluronic acid-coated liposome-protamine-DNA complex for pulmonary inhalation. Int J Pharm. 2020;583:119338. DOI: 10.1016/j.ijpharm.2020.119338

[61] Ly A, Carrigy NB, Wang H, Harrison M, Sauvageau D, Martin AR, Vehring R, Finlay WH. Atmospheric spray freeze drying of sugar solution with phage D29. Front Microbiol. 2019;10:488. DOI: 10.3389/fmicb.2019.00488

[62] Qiu Y, Liao Q, Chow MYT, Lam JKW. Intratracheal sdministration of dry powder formulation in mice. J Vis Exp JoVE. 2020. DOI: 10.3791/61469

[63] Chang RYK, Chen K, Wang J, Wallin M, Britton W, Morales S, Kutter E, Li J, Chan HK. Anti-Pseudomonal activity of phage PEV20 in a dry powder formulation — A proof-of-principle study in a murine lung infection model. Antimicrob Agents Chemother. 2017. DOI: 10.1128/AAC.01714-17

[64] Hoe S, Boraey MA, Ivey JW, Finlay WH, Vehring R. Manufacturing and device options for the delivery of biotherapeutics. J Aerosol Med Pulm Drug Deliv. 2013;27:315-328. DOI: 10.1089/jamp.2013.1090

[65] Terzano C. Pressurized metered dose inhalers and add-on devices. Pulm Pharmacol Ther. 2001;14:351-366. DOI: 10.1006/pupt.2001.0273

[66] Hochrainer D, Hölz H, Kreher C, Scaffidi L, Spallek M, Wachtel H. Comparison of the aerosol velocity and spray duration of Respimat Soft Mist

inhaler and pressurized metered dose inhalers. J Aerosol Med Off J Int Soc Aerosols Med. 2005;18:273-282. DOI: 10.1089/jam.2005.18.273

[67] Hyman P, Abedon ST. Chapter 7 - Bacteriophage host range and bacterial resistance. Adv. Appl. Microbiol., vol. 70, Academic Press; 2010, p. 217-48. DOI: 10.1016/S0065-2164(10)70007-1

[68] Himmelweit F. Combined action of penicillin and bacteriophage on Staphylococci. Lancet. 1945:104-105

[69] Letrado P, Corsini B, Díez-MartínezR, BustamanteN, YusteJE, García P. Bactericidal synergism between antibiotics and phage endolysin Cpl-711 to kill multidrug-resistant pneumococcus. Future Microbiol. 2018;13:1215-1223. DOI: 10.2217/fmb-2018-0077

[70] Akturk E, Oliveira H, SantosSB, CostaS, KuyumcuS, MeloLDR, Azeredo J. Synergistic action of phage and antibiotics: Parameters to enhance the killing efficacy against mono and dual-species biofilms. antibiotics. 2019;8:103. DOI: 10.3390/antibiotics8030103

[71] Oechslin F, Piccardi P, Mancini S, Gabard J, Moreillon P, Entenza JM, Resch G, Que YA. Synergistic interaction between phage therapy and sntibiotics clears Pseudomonas aeruginosa infection in endocarditis and reduces virulence. J Infect Dis. 2017;215:703-712. DOI: 10.1093/infdis/jiw632

[72] Knezevic P, Curcin S, Aleksic V, Petrusic M, Vlaski L. Phage-antibiotic synergism: a possible approach to combatting Pseudomonas aeruginosa. Res Microbiol. 2013;164:55-60. DOI: 10.1016/j.resmic.2012.08.008

[73] Ryan EM, Alkawareek MY, Donnelly RF, Gilmore BF. Synergistic phage-antibiotic combinations for the control of Escherichia coli biofilms in vitro. FEMS Immunol Med Microbiol. 2012;65:395-398. DOI: 10.1111/j.1574-695X.2012.00977.x

[74] Vouillamoz J, Entenza JM, Giddey M, Fischetti VA, Moreillon P, Resch G. Bactericidal synergism between daptomycin and the phage lysin Cpl-1 in a mouse model of pneumococcal bacteraemia. Int J Antimicrob Agents. 2013;42:416-421. DOI: 10.1016/j.ijantimicag.2013.06.020

[75] Kirby AE. Synergistic action of gentamicin and bacteriophage in a continuous culture population of Staphylococcus aureus. Plos One. 2012;7:e51017. DOI: 10.1371/journal.pone.0051017

[76] Nouraldin AAM, Baddour MM, Harfoush RAH, Essa SAM. Bacteriophage-antibiotic synergism to control planktonic and biofilm producing clinical isolates of Pseudomonas aeruginosa. Alex J Med. 2016;52:99-105-99-105. DOI: 10.4314/bafm.v52i2

[77] Coulter LB, McLean RJC, Rohde RE, Aron GM. Effect of bacteriophage infection in combination with tobramycin on the emergence of resistance in Escherichia coli and Pseudomonas aeruginosa biofilms. Viruses. 2014;6:3778-3786. DOI: 10.3390/v6103778

[78] Torres-Barceló C, Arias-Sánchez FI, Vasse M, Ramsayer J, Kaltz O, Hochberg ME. A window of opportunity to control the bacterial pathogen Pseudomonas aeruginosa combining antibiotics and phages. PloS One. 2014;9:e106628. DOI: 10.1371/journal.pone.0106628

[79] Verma V, Harjai K, Chhibber S. Restricting ciprofloxacin-induced resistant variant formation in biofilm of Klebsiella pneumoniae B5055 by complementary bacteriophage

treatment. J Antimicrob Chemother. 2009;64:1212-1218. DOI: 10.1093/jac/dkp360

[80] Comeau AM, Tétart F, Trojet SN, Prère M-F, Krisch HM. Phage-Antibiotic Synergy (PAS): beta-lactam and quinolone antibiotics stimulate virulent phage growth. PloS One. 2007;2:e799. DOI: 10.1371/journal.pone.0000799

[81] Górski A, Międzybrodzki R, Borysowski J, editors. Phage Therapy: A Practical Approach. Cham: Springer International Publishing; 2019. DOI: 10.1007/978-3-030-26736-0

[82] Chan BK, Sistrom M, Wertz JE, Kortright KE, Narayan D, Turner PE. Phage selection restores antibiotic sensitivity in MDR Pseudomonas aeruginosa. Sci Rep. 2016;6:26717. DOI: 10.1038/srep26717

[83] Torres-Barceló C, Hochberg ME. Evolutionary rationale for phages as complements of antibiotics. Trends Microbiol. 2016;24:249-256. DOI: 10.1016/j.tim.2015.12.011

[84] Gordillo Altamirano F, Forsyth JH, Patwa R, Kostoulias X, Trim M, Subedi D, Archer SK, Morris FC, Oliveira C, Kielty L, Korneev D, O'Bryan MK, Lithgow TJ, Peleg AY, Barr JJ. Bacteriophage-resistant Acinetobacter baumannii are resensitized to antimicrobials. Nat Microbiol. 2021;6:157-161. DOI: 10.1038/s41564-020-00830-7

[85] Chaudhry WN, Concepción-Acevedo J, Park T, Andleeb S, Bull JJ, Levin BR. Synergy and order effects of antibiotics and phages in killing Pseudomonas aeruginosa biofilms. PloS One. 2017;12:e0168615. DOI: 10.1371/journal.pone.0168615

[86] Kumaran D, Taha M, Yi Q, Ramirez-Arcos S, Diallo J-S, Carli A, Abdelbary H. Does treatment order matter? Investigating the ability of bacteriophage to augment antibiotic activity against Staphylococcus aureus biofilms. Front Microbiol. 2018;9. DOI: 10.3389/fmicb.2018.00127

[87] Liu CG, Green SI, Min L, Clark JR, Salazar KC, Terwilliger AL, Kaplan HB, Trautner BW, Ramig RF, Maresso AW. Phage-antibiotic synergy is driven by a unique combination of antibacterial mechanism of action and stoichiometry. MBio. 2020;11. DOI: 10.1128/mBio.01462-20

[88] Rodriguez-Gonzalez RA, Leung CY, Chan BK, Turner PE, Weitz JS. Quantitative models of phage-antibiotic combination therapy. MSystems. 2020;5. DOI: 10.1128/mSystems.00756-19

[89] Lin Y, Chang RYK, Britton WJ, Morales S, Kutter E, Chan H-K. Synergy of nebulized phage PEV20 and ciprofloxacin combination against Pseudomonas aeruginosa. Int J Pharm. 2018;551:158-165. DOI: 10.1016/j.ijpharm.2018.09.024

[90] Lin Y, Chang RYK, Britton WJ, Morales S, Kutter E, Li J, Chan HK. Inhalable combination powder formulations of phage and ciprofloxacin for P. aeruginosa respiratory infections. Eur J Pharm Biopharm. 2019;142:543-552. DOI: 10.1016/j.ejpb.2019.08.004

[91] Lin Y, Quan D, Chang RYK, Chow MYT, Wang Y, Li M, Morales S, Britton WJ, Kutter E, Li J, Chan HK. Synergistic activity of phage PEV20-ciprofloxacin combination powder formulation—A proof-of-principle study in a P. aeruginosa lung infection model. Eur J Pharm Biopharm. 2021;158:166-171. DOI: 10.1016/j.ejpb.2020.11.019

[92] Lin Y, Yoon Kyung Chang R, Britton WJ, Morales S, Kutter E, Li J, Chan HK. Storage stability of phage-ciprofloxacin combination powders against Pseudomonas aeruginosa

respiratory infections. Int J Pharm. 2020;591:119952. DOI: 10.1016/j.ijpharm.2020.119952

[93] Górski A, Borysowski J, Międzybrodzki R. Phage therapy: Towards a successful clinical trial. Antibiot Basel Switz. 2020;9. DOI: 10.3390/antibiotics9110827

[94] Singla S, Harjai K, Katare OP, Chhibber S. Encapsulation of bacteriophage in liposome accentuates its entry in to macrophage and shields it from neutralizing antibodies. PloS One. 2016;11:e0153777. https://doi.org/10.1371/journal.pone.0153777

[95] Shan J, Ramachandran A, Thanki AM, Vukusic FBI, Barylski J, Clokie MRJ. Bacteriophages are more virulent to bacteria with human cells than they are in bacterial culture; insights from HT-29 cells. Sci Rep. 2018;8:5091. DOI: 10.1038/s41598-018-23418-y

[96] Núñez-Sánchez MA, Colom J, Walsh L, Buttimer C, Bolocan AS, Pang R, Gahan CGM, Hill C. Characterizing phage-host interactions in a simplified human intestinal barrier model. Microorganisms. 2020;8:1374. DOI: 10.3390/microorganisms8091374

[97] Bichet MC, Chin WH, Richards W, Lin Y, Avellaneda-Franco L, Hernandez CA, Oddo A, Chernyavskiy O, Hilsenstein V, Neild A, Li J, Voelcker NH, Patwa R, Barr JJ. Bacteriophage uptake by Eukaryotic cell layers represents a major sink for phages during therapy. bioRix Posted September 8, 2020. DOI: 10.1101/2020.09.07.286716.

Chapter 10

Challenges of Phage Therapy as a Strategic Tool for the Control of *Salmonella Kentucky* and Repertoire of Antibiotic Resistance Genes in Africa

Igomu Elayoni Emmanuel

Abstract

Salmonella Kentucky ST198 (*S. Kentucky* ST198) is the most ubiquitous multidrug resistant (MDR) strain posing the greatest threat to public health, livestock and food industry in Africa. The reinvention of bacteriophage (Phage) as a non-antibiotic alternative only gives a glimmer of hope in the control of MDR strains of *Salmonellae*. *S. Kentucky* ST198 posses' chromosomal and plasmid factors capable of been co-opted into phage mediated transduction and co-transduction of antibiotic resistance genes (ARGs) as well as cross-serovar transduction of ARGs. Phage DT104, DT120 and P-22 like prophages like PDT17 and ES18 together have been shown to be capable of transducing and co-transducing the classical ACSSuT resistance phenotype identified in most *S. Kentucky* ST198 strain on the continent. Also, the institution of fluoroquinolones and third generation cephalosporin for salmonellosis treatment in animals or human infected by *S. Kentucky* ST198 strain resistant to these drugs can induce *Salmonella* phage transduction of kanamycin between different *Salmonella* serovars if present. This review highlights possible risk associated with the use of known *Salmonella* phages in the control of *S. Kentucky* ST198 and the need for chromosomal and plasmid tracking of genes prior to the institution of phage therapy on the continent.

Keywords: Bacteriophages, *Salmonella Kentucky* ST198, DT104, transduction, ARGs, *Salmonella* conjugative plasmids, Africa

1. Introduction

Bacteriophages (here in after called phages are viruses that can infect a bacteria and replicate within it) are completely alien to the routine therapeutic regimens in both veterinary and human medical practices in Africa, and where phage therapies have been instituted they are mainly experimental. Phage therapy has shown to be an ecologically sustainable tool in the control of bacterial infection; scientific researches places phages to be superiorly bactericidal specific, efficacious and cost effective when compared to antibiotics and interestingly it has been proven to inhibit biofilm formation in pathogenic bacteria [1–3], customarily the production of biofilms by bacterial cells significantly increases their resistance to antimicrobials as compared to what is normally seen by the same cells being planktonic [4]. In Africa, the indiscriminate use of antimicrobials for treatment of salmonellosis in both human medical and veterinary practices has allowed for the proliferation of multidrug resistant (MDR) determinants and the sharing of antibiotic resistant genes (ARGs) between serovars of *Salmonellae* and other bacteria population, and on a continent where Poverty, human immunodeficiency virus/acquired immune deficiency syndrome (HIV/AIDS), Malaria, Tuberculosis and other immuno-compromising diseases are prevalent, the impact of MDR salmonellosis is severe [5]. It has become pertinent that alternative strategies for control of salmonellosis and *Salmonella* associated infections be adopted especially where MDR strains of *Salmonellae* persist. *Salmonella Kentucky* (*S. Kentucky*) is amongst the most ubiquitous *Salmonella* serovar identified on the African continent in the present decade and MDR strains pose significant health risk and a threat to the livestock production, livestock trade and food industry [5]. Several MDR strains have been isolated in most regions of the continent and the abusive use of antibiotics has only enhanced its mutative MDR tendencies and acquisition of ARGs between serovars and other non-genera of *Salmonella*. The institution of phages in the treatment of salmonellosis has shown promising results because of their very low transduction frequencies in the transmission of ARGs of *Salmonella* spp. [6], they exist everywhere in the environment and are natural, economically sustainable, nontoxic and some phages have shown broad activity against numerous serovars of MDR *Salmonella* spp. [7, 8]. Felix-O1 and SE13 are examples of *Salmonella* phages with broad serovar capacities; Felix-O1, a virulent phage was proven to infect 98.2% of all *Salmonella* strain and SE13 was capable of lysing 83.6% of *Salmonella* strain it was tested with [7, 8]. While researches on the use of phages for the treatments of *S. Kentucky* in Africa are scarce, [9] reported an effective use of phage in the reduction of *S. Kentucky* colonization in different broiler farms in Egypt. Phage–host interactions through the mechanism of horizontal gene transfer have contributed significantly to genetic flux vastly responsible for the acquisition and dissemination of important bacterial phenotypes, such as enhanced colonization of the human or animal gut epithelium, AMR and toxin production [10, 11]. Thus, the identification and careful selection of phages devoid of genetic elements that could pose risk to human and animal health is critical to biocontrol applications [12]. This review proposes to highlight challenges that may arise in the institution of phages as a strategic non-antibiotic tool for the control *S. Kentucky* and repertoire of its ARGs without prior studies on their genetic make-up.

2. Antibiotic resistant gene

Antibiotic resistance genes (ARGs) are an emerging public health contaminant, posing a potential global health risk. A major factor contributing to the increased environmental burden of ARGs is the rise in intensive livestock farming [13]. The World Health Organization (WHO) defines antimicrobial resistance (AMR) as "an increase in the minimum inhibitory concentration of a compound for a previously sensitive strain" [14]. Human beings consistently use large amounts of antibiotic in the human medical contexts as well as for growth factors and prophylaxis in agriculture and livestock, culminating in the contamination of environmental microbial communities. Unfortunately, even when pathogenic bacteria are the specific targets of antibiotic use, hundreds of non-pathogenic bacteria species are affected [15]. Thus, antibiotics are present in microbial communities, not only as a result of the natural lifecycle of microorganisms but also to the usage of these drugs in agriculture, food industry, livestock and human health [16]. The presence of antibiotic resistance genes in environmental bacteria may be responsible for different mechanisms employed to overcome the natural antibiotics present in the environment. Recently this gene pool has been named the 'resistome', and its components can be mobilized into the microbial community affecting humans because of the participation of genetic platforms that efficiently facilitate the mobilization, transmission and maintenance of these resistance genes. Evidence for this transference has been suggested and or demonstrated using cutting-edge research techniques with newly identified widespread genes in multidrug-resistant bacteria [17]. These resistance genes include those responsible for plasmid-mediated efflux pumps conferring low-level fluoroquinolone resistance (*qepA*), ribosomal methylases affecting aminoglycosides (*armA, rtmB*) and methyltransferases affecting linezolid (*cfr*) all of which have been associated with antibiotic-producing bacteria. Recently, resistance genes whose ancestors have been identified in environmental isolates that are not recognized as antibiotic producers have also been detected. These include the *qnr* and the *blaCTX* genes compromising the activity of fluoroquinolones and extended-spectrum cephalosporins, respectively [17]. Bacteria can express antibiotic resistance through chromosomal mutations or via the acquisition of genetic material through horizontal gene transfer from other bacteria or the environment. Acquisition of genetic material via horizontal gene transfer is largely driven by mobile genetic elements (MGEs), such as plasmids, transposons or bacteriophages, which play a critical role in the evolution and ecology of bacterial communities by controlling the intra-species and interspecies exchange of genetic information [18]. While the transfer of these MGEs usually occur through transformation, transduction, or conjugation, conjugation is mostly considered the most efficient mechanism employed for the exchange of genetic material among bacteria [19]. The ease of acquisition and spread of ARGs by bacteria via conjugation is frequently through conjugative plasmids and transposons, and the contribution of these elements to antibiotic resistance pool has been extensively studied in hospital, community, agricultural and environmental settings [15–17, 20, 21], but very little is known about the role of bacteriophages as vehicles for ARGs in environmental settings. Recent findings based on cutting-edge genomic technologies suggest that, in these settings, bacteriophages play a more important role in the mobilization of ARGs than previously documented [22].

3. Phage transduction: primary mechanism for the transfer of ARGs

Intensive studies of the mechanisms for horizontal gene transfer responsible for the increased spread of antibiotic resistance to foodborne bacterial pathogens have been undertaken; Conjugation, transformation, and transduction are the fundamental mechanisms by which dissemination of ARGs occurs [23]. Transduction is primarily the horizontal gene transfer mechanisms employed by most phages, and recent findings have shown phage-mediated transduction to be a significant driver in the dissemination of ARGs [24]. The concept that phage mediated transduction is a major driver of horizontal transfer of ARGs between foodborne pathogens, as well as from the environment to animals and humans, is increasingly been recognized. Phages are recognized as the most abundant organism in the biosphere, and are found in every environment regardless of their diversities, including oceans, lakes, soil, urban sewage, potable and well water, plant and animal microbial communities [25]. ARGs are often found on various MGEs, and are readily transferred horizontally by phage transduction [24]. Phages infect bacteria and either incorporate their viral genome into the host genome, replicating as part of the host (lysogenic cycle), or replicate inside the host cell before releasing new phage particles (lytic cycle) [22, 26]. Phages can be either virulent or temperate. The mechanism of transduction has been vastly described in virulent phages (defined by their capacity to undergo lytic cycles). Following bacterial infection, there is an immediate induction of phage particles formation and lysis of the host cell but virulent phages do not integrate their DNA into the host chromosome. Temperate phages (known to undergo lysogenic cycle), integrate their DNA into the host chromosome and the prophage may remain dormant in the host until other factors like stress induces the excision of the phage from the chromosome leading to subsequent formation of phage particles and lysis of the host cell. Some phages can also adopt a pseudolysogenic state under unfavourable growth condition. In this state, their genome does not degrade but rather exist within the host cytoplasm as a plasmid and during bacterial cell division becomes incorporated into just one daughter cell [26]. Genetic materials are transferred between hosts either by generalized or specialized transduction. Virulent and temperate phages can undergo generalized transduction, here, bacterial DNA fragments are randomly packaged into the phage capsid during their lytic cycle forming a "transducing particle". These "offspring" phages do not contain phage genes, and only the capsid has a viral origin. Despite this, the transducing particle is capable of injecting the bacterial genes into a susceptible recipient cell, which can subsequently be incorporated into the host genome by recombination [5, 22, 24]. Specialized transduction is restrictive to temperate phages and results in the packaging of bacterial DNA into phages at a higher frequency; temperate phages insert their genomes into a specific region of the host chromosome. An inaccurate excision of the prophage may lead to the capture of the flanking genes adjacent to the phage integration point. If capsids carrying the rearranged phage genome with these foreign genes infect other bacteria and integrate into the host chromosome, transduction of the acquired genes will be achieved. However, the probability that the transferred genes are antibiotic resistance-related is relatively low [5].

4. Phage transduction of ARGs in *Salmonellae*

Salmonella phages have been extensively used in molecular biology for the introduction of foreign genes by generalized and specialized transduction. P-22, a well-known phage is a classical example, other P-22 like prophages ST104 or PDT17, harboured within DT104 phage type have been hypothesized to facilitate horizontal transfer of the penta-resistance genes [27, 28]. The penta-resistance genes in phage type DT104 are clustered on a 43-kb *Salmonella* genomic island-1 (SGI1), which is flanked by two type I integrons [29]. *Salmonella* genomic island 1 (SGI1) is an integrative mobilizable element that harbours a multidrug resistance (MDR) gene cluster. A research undertaken by [27] asserted that ES18 and PDT17, also a P-22 like phage, following release from DT104 could transduce ARGs. Their findings further demonstrated the transduction of *cam* and *amp* by phage PDT17 and *amp*, *cam*, and *tet*, which confer resistance to ampicillin, chloramphenicol, and tetracycline, respectively, by ES18 from a donor DT104 strain into a DT104 recipient strain lacking these resistance genes. Phage ES18 also co-transduces selected ARGs of the 71 *tet* transductants and of the 145 *cam* transductants. Interestingly, in 14 of 16 transductants, it was noticed that phage E18 could co- transduce *sul* and *str*, genes involved in resistance to sulphonamides and streptomycin, respectively, together with *amp*, *cam*, and *tet* to create the ACSSuT (ampicillin, chloramphenicol, streptomycin, sulphonamides, and tetracycline) resistance phenotype [27]. This co-transduction likely occurs because *amp* and *str* are situated on the integrons flanking SGI1, and the phage likely packages the SGI1 and its flanking integrons [30–32]. P-22 phage has also been identified within DT120 isolates shown to be capable of generalized transduction and possess the ACSSuT resistant strain [33]. See [33] reported that carbadox, a veterinary antibacterial that posses' mutagenic and carcinogenic capabilities induced phage transduction in DT104 and DT120. Furthermore the absence of transduction in DT104 strain which had its P-22 like prophage deleted following induction with carbadox suggests that P-22 like prophages are responsible for generalized transduction. Thus; transduction and co-transduction by P22-like prophages of ARGs co-located within SGI1 in multidrug-resistant *Salmonellae* strains is a common phenomenon. Also, genome scanning proved that P22-like prophages were common in 18 *Salmonella* serovars implying that generalized transduction may be greatly underestimated [33].

5. Transduction of Ciprofloxacin and cephalosporins genes

Ciprofloxacin a fluoroquinolone and third generation cephalosporin are the drugs of choice in the treatment of invasive *Salmonella* infections [34–36]. Resistance of *Salmonella* to ciprofloxacin is due mainly to double mutations in *gyrA* and a single mutation in *parC* genes. In addition, *oqxAB* operon is suggested to be responsible for the increase in resistance observed in clinical *Salmonella* strains [37]. It was observed by [38], that Ciprofloxacin, enrofloxacin and danofloxacin induced *Salmonella* phage DT104 and DT102 transfer of a native kanamycin resistance plasmid to a strain of *Salmonella Typhimurium* by generalized transduction. Resistance to cephalosporin is mainly due to extended spectrum beta-lactamases (ESBLs), such as TEM-, SHV-, and CTX-M, or plasmid mediated AmpC β-lactamases (pAmpCs), such as CMY, encoded on transmissible conjugative plasmids [39–41], or be transferred by generalized transduction. Phage P24, induced from an isolate of *S. Typhimurium*, was propagated on a multidrug resistant strain of *S. Heidelberg* (S25). Thus, when the MDR S25 harbouring phage P24 was used as transduction donor to transfer ESBL and tetracycline resistance genes to a recipient *S. Typhimurium* isolate. PCR confirmed the presence of *bla*CMY-2, *tet*(A), and *tet*(B) in various *S. Typhimurium* transductants. Although the tetracycline genes were not co-transduced with *bla*CMY-2, their transduction frequency was equivalent, indicating generalized transduction and evidently reporting the transfer of ARGs by phage-mediated transduction between different *Salmonella* serovars. This finding likely expresses that cross-serovar transduction occurs frequently because phages can bind to various surface protein receptors on different species and serovars [42]. The LPS, FliC, OmpC, OmpF, OmpA, are examples of phage receptors present in *Salmonella* [43]. In the previous study [42] it was observed that 13 inducible phages recovered from 31 *Salmonella* serovars were capable of propagating on two or more *Salmonella* serovars including those often responsible for foodborne outbreaks such as *S. Heidelberg, S. Enteritidis, S. Typhimurium* and *S. Kentucky*. Finally, the findings of [42] demonstrate the spread of antibiotic resistance in *Salmonellae* by phage mediated transduction.

6. Transduction of R-factor genes

R-factors which are a group of conjugative plasmids that harbour one or more antibiotic resistance determinants and represents another form of MGE that can be transferred horizontally by phage mediated transduction [24]. Conjugative plasmids are also self-transmissible, affording them the capacities to increase the spread of ARGs. The origin of transfer (*ori*T), MOB genes, and the mate-pair formation (MPF) genes are the essential components for conjugation [44, 45]. In order for conjugation to occur, a protein complex called a 'relaxasome' responsible for processing plasmid DNA to prepare it for transfer must form at the *ori*T [46–48] and the mechanism for R-factor-phage acquisition and propagation of ARGs may be random. R factors in close proximity to P22-like prophages could be integrated into the head of the assembling phage during induction from its host, thus contributing to the spread of ARGs within bacteria capable of causing foodborne illnesses, in the intestinal flora of livestock and in the environment [24].

7. Virulent factors of *S. Kentucky* ST198 that may potentiate phage-mediated transduction of ARGs

Salmonella Kentucky ST198 is a global contaminant and an emerging risk for foodborne illness, although first identified in Egypt it has now been isolated in several countries across the different regions in Africa [5, 49], with reservoirs in various animals and food [49–54]. Successes have been recorded with the institution of phages in controlling the spread of MDR *Salmonella Kentucky* [9, 55]; these findings however rarely discuss the tendencies of phage-host mediated propagation of ARGs or other MDR determinants. *S. Kentucky* ST198 belongs to a single lineage, which is predicted to emerged circa 1989 following the acquisition of the AMR-associated *Salmonella* genomic island (SGI) 1 (variant SGI1-K), that confers resistance to ampicillin, streptomycin, gentamicin, sulfamethoxazole and tetracycline [56]. This MDR *Salmonella Kentucky* clone has undergone substitution mutations in the quinolone-resistance-determining regions (QRDRs) of DNA gyrase (gyrA) and DNA topoisomerase IV (parC) genes, such that most strains carry three QRDR mutations which together confer resistance to ciprofloxacin. Its molecular characterization further shows a chromosomal genomic island carrying resistance genes that confer resistance to β-lactam antibiotics, carbapenems, quinolones, aminoglycosides, co-trimoxazole (trimethoprim-sulfamethoxazole), and to Azithromycin. Extended-spectrum cephalosporins (ESCs) resistance has also been associated with *S. Kentucky* ST198 [57–60]. Genetic basis for this resistance showed an extended-spectrum b-lactamase (ESBL) [61]. The aforementioned resistant properties evidently can allow for the transfer of native kanamycin resistance plasmid to strains of *S. Typhimurium* or other *Salmonella serovar* by generalized transduction as treatment with these antibiotics as reported by [38] can induce *Salmonella* phage DT104 and DT102 transmission of a native kanamycin resistance plasmid and other ARGs between serovars of *Salmonella* by generalized transduction. *S. Kentucky* also exhibits an extensive MDR pattern with diverse resistance profile cutting across human, environmental and poultry micro biomes [57]. A penta-resistant profile (SSuTCipNa) was observed in *S. Kentucky* from human, environmental and poultry samples with a deca-resistant profile, ACKSSuSxTAmcCipNa in poultry [57, 58]. *Salmonella* Phages ES18, PDT17, DT104, DT120 and other P22-like prophages like ST104 or PDT17 haboured within DT104 have been proven to be participatory in the transduction and co-transduction of genes for ACSSuT resistance phenotype [33], making *S. Kentucky* ST198 a luxurious menu for the transduction of these genes complemented by other factors that may helps in phage-mediated transduction of ARGs between serovars of *Salmonella* and other enterobacteriaceae. Several conjugative plasmids have also been detected in *S. Kentucky* ST198; IncA/C conjugative plasmids have been isolated in *S. Kentucky* ST198 that contain up to 10 ARGs for more than five classes of antibiotics. The most common ARGs carried by IncA/C are *str*AB (aminoglycosides), *sul*2 (sulfonamides), *tet*AR (tetracycline), *bla*CMY-2 (β-lactams), *floR* (chloramphenicols) and *bla*CTX-M-25 (cephalosporin). Other genes for resistance to aminoglycosides, tetracyclines, trimethoprim, chloramphenicols, and have also been identified [49, 62–64]. *S. Kentucky* ST198 also contains an IncF plasmid [65]. IncF plasmids can carry multiple types of replicon associated genes, such as FIA, FII, or FIB [66]. IncF plasmids have been observed to contain ARGs, exhibiting resistance to fluoroquinolone [67], they have also been associated with *str*AB, *tet*A, *tet*C, *tet*D, *aph*A (aminoglycosides), and *sul*2 (sulphonamides) resistance [68, 69]. Another plasmid of importance carried by *S. Kentucky* ST198 is the IncHI plasmid and has been associated with *qnr* genes (fluoroquinolones) and ESBL genes [70]. The integration of one of more of these conjugative plasmids that may be in close proximity to P22-like prophages would facilitate their packaging into the core of the assembling phage during induction from its host, thus contributing to the spread of antibiotic resistance between generic and non-generic bacteria in the intestinal flora of livestock and human and in the environment causing foodborne illnesses and outbreaks. Although the mode of acquisition of Plasmids ARGSs in *Salmonella* may seem random, their proliferation in a population is usually not random. Consequently, surveillance is a necessary tool, not just for *Salmonella* and other important human and animal pathogens, but for the plasmids they carry. Therefore, the tracking of plasmids and the genes they carry would allow for a better understanding of co-selection of ARGs and the associations of plasmids with *Salmonella* serotypes [71]. Finally since *Salmonellae* phages can bind to several protein receptors in *Salmonellae* and other members of the enterobacteriaceae family thereby permitting cross-serovar and inter-specie transduction of ARGs [43], it has become necessary that measures or protocols that can hinder such developments be adopted in order to forestall the spread of *Salmonellae* associated foodborne outbreaks.

8. Conclusion

The renewed and profound interest in phage therapy as a non-antibiotic measure for combating MDR strains of *Salmonellae* is a testament to their efficacy, but clearly *Salmonella Kentucky* ST198 posse's virulent factors that can potentiate phage-mediated cross-serovar transduction and co-transduction of ARGs and MDR determinants, therefore investigative laboratory protocol should therefore be sought to identify these determinants prior to the institution of phages in the treatment of non-repressive salmonellosis.

Conflict of interest

Author declares none.

Author details

Igomu Elayoni Emmanuel
Bacterial Vaccine Production Division, National Veterinary Research Institute, P. M. B. 01 Vom, Plateau State, Nigeria

*Address all correspondence to: elayonigomu@gmail.com

References

[1] Bhardwaj SB, Mehta M, Sood S, Sharma J. Isolation of a Novel Phage and Targeting Biofilms of Drug-Resistant Oral *Enterococci*. Journal of global infectious diseases. 2020; 12(1): 11-15. https://doi.org/10.4103/jgid.jgid_110_19.

[2] Principi N, Silvestri E, Esposito S. Advantages and Limitations of Bacteriophages for the Treatment of Bacterial Infections. Frontiers in Pharmacology. 2019; 10(513). doi: 10.3389/fphar.2019.00513

[3] Wei S, Chelliah R, Rubab M, Oh D-H, Uddin MJ, Ahn J. Bacteriophages as Potential Tools for Detection and Control of *Salmonella* spp. in Food Systems. Microorganisms. 2019; 7 (570): 1-22.

[4] Mah TFC, O'Toole GA. Mechanism of biofilm resistance to antimicrobial agents. Trends in Microbiology. 2001; 9(1): 34-39.

[5] Igomu EE. *Salmonella Kentucky*: prevalence and challenges in Nigeria and the Africa continent. African Journal of Clinical and Experimental Microbiology. 2020; 21(4): 272-283.

[6] Torres-Barceló C. The disparate effects of bacteriophages on antibiotic - resistant bacteria. Emerging Microbes and Infections. 2018; 7(1): 1-12.

[7] Fong K, Tremblay DM, Delaquis P, Goodridge L, Levesque RC, Moineau S, Suttle CA, Wang S. Diversity and Host Specificity Revealed by Biological Characterization and Whole Genome Sequencing of Bacteriophages Infecting *Salmonella enterica*. Viruses. 2019; 11(854): 1 -19.

[8] Mohamed A, Taha O, El-Sherif HM, Connerton PL, Hooton SPT, Bassim ND, Connerton IF, El-Shibiny A. Bacteriophage ZCSE2 is a Potent Antimicrobial against *Salmonella enterica* Serovars: Ultrastructure, Genomics and Efficacy. Viruses. 2020; 12(424): 1- 18.

[9] Sorour HK, Gaber AF, Hosny RA. Evaluation of the efficiency of using *Salmonella Kentucky* and *Escherichia coli* O119 bacteriophages in the treatment and prevention of salmonellosis and colibacillosis in broiler chickens. Letters in Applied Microbiology. 2020; DOI: 10.1111/lam.13347

[10] Turner D, Ackermann HW, Kropinski AM, Lavigne R, Sutton JM, Reynolds DM. Comparative analysis of 37 *Acinetobacter* bacteriophages. Viruses. 2018; 10:5. doi: 10.3390/v10010005

[11] Seed KD. Battling phages: How bacteria defend against viral attack. PLoS Pathogens 2015; 11, e1004847. https://doi.org/10.1371/journal.ppat.1004847.

[12] Goodridge L, Fong K, Wang S, Delaquis P. Bacteriophage-based weapons for the war against foodborne pathogens. Current Opinion in Food Science. 2018; 20: 69-75.

[13] Zhu Y-G, Johnson TA, Su J-Q, Qiaob M, Guob G-X, Stedtfeld RD, Hashsham SA, Tiedje JM. Diverse and abundant antibiotic resistance genes in Chinese swine farms. PNAS. 2013; 110 (9): 3435-3440.

[14] World Health Organization [WHO]. *Antimicrobial Resistance (factsheet no194)*. Geneva: World Health Organization 2013.

[15] Escudeiro P, Pothier J, Dionisio F, Nogueira T. Antibiotic resistance gene diversity and virulence gene diversity are correlated in human gut and environmental microbiomes. mSphere. 2019; 4 (3): 1-9.

[16] Castanon J. History of the use of antibiotic as growth promoters

in European poultry feeds. Poultry Science. 2007; 86: 2466-2471.

[17] Canton R. Antibiotic resistance genes from the environment: a perspective through newly identified antibiotic resistance mechanisms in the clinical setting. Clinical Microbiology and Infection. 2009; 15(1): 20-25.

[18] Frost LS, Leplae R, Summers AO, Toussaint A. Mobile genetic elements: the agents of open source evolution. Nature Review Microbiology. 2005; 3:722-732.

[19] Courvalin P. Transfer of antibiotic resistance genes between Gram-positive and Gram negative bacteria. Antimicrobial Agents and Chemotherapy. 1994; 38:1447-1451.

[20] Hardiman CA, Weingarten RA, Conlan S, Khil P, Dekker JP, Mathers AJ. Horizontal transfer of carbapenemase-encoding plasmids and comparison with hospital epidemiology data. Antimicrobial Agents and Chemotherapy. 2016; 60(8): 4910-4919.

[21] Chen S, Zhao S, White DG, Schroeder CM, Lu R, Yang, H. Characterization of multiple-antimicrobial-resistant *Salmonella* serovars isolated from retail meats. Applied Environmental Microbiology. 2004; 70: 1-7.

[22] Balcazar JL. How do bacteriophages promote antibiotic resistance in the environment? Clinical Microbiology and Infection. 2018; 24: 447 – 449

[23] von Wintersdorff CJ, Penders J, Van Niekerk JM, Mills ND, Majumder S, Van Alphen LB. Dissemination of antimicrobial resistance in microbial ecosystems through horizontal gene transfer. Frontiers in Microbiology. 2016; 7:173. doi:10.3389/fmicb.2016.00173

[24] Colavecchio A, Cadieux B, Lo A, Goodridge LD. Bacteriophages Contribute to the Spread of Antibiotic Resistance Genes among Foodborne Pathogens of the Enterobacteriaceae Family – A Review. Frontiers in Microbiology. 2017; 8 (1108):1-13.

[25] Clokie MR, Millard AD, Letarov AV, Heaphy S. Phages in nature. Bacteriophage. 2011; 1: 31-45.

[26] Feiner R, Argov T, Rabinovich L, Sigal N, Borovok I, Herskovits AA. A new Perspective on lysogeny: prophages as active regulatory switches of bacteria. Nature Review Microbiology. 2015; 13: 641-650.

[27] Schmieger H, Schicklmaier P. Transduction of multiple drug resistance of *Salmonella enterica* serovar *Typhimurium* DT104. FEMS Microbiology Letter. 1999; 170: 251-256.

[28] Tanaka K, Nishimori K, Makino S-I, Nishimori T, Kanno T, Ishihara R. Molecular characterization of a prophage of *Salmonella enterica* serotype *Typhimurium* DT104. Journal of Clinical Microbiology. 2004; 42: 1807-1812.

[29] Boyd D, Cloeckaert A, Chaslus-Dancla E, Mulvey MR. Characterization of Variant *Salmonella* genomic island 1 multidrug resistance regions from serovars *Typhimurium* DT104 and *Agona*. Antimicrobial Agents and Chemotherapy. 2002; 46: 1714-1722.

[30] Ridley A, Threlfall EJ. Molecular epidemiology of antibiotic resistance genes in multiresistant epidemic *Salmonella Typhimurium* DT 104. Microbial Drug Resistance. 1998; 4: 113-118.

[31] Sandvang D, Aarestrup FM, Jensen LB. Characterisation of integrons and Antibiotic resistance genes in Danish multiresistant *Salmonella enterica Typhimurium* DT104. FEMS Microbiology Letter. 1998; 160: 37-41.

[32] Mulvey MR, Boyd DA, Olson AB, Doublet B, Cloeckaert A. The genetics

of *Salmonella* genomic island 1. Microbes of Infection. 2006; 8: 1915-1922

[33] Bearson BL, Allen HK, Brunelle BW, Lee IS, Casjens SR, Stanton TB. The Agricultural antibiotic carbadox induces phage-mediated gene transfer in *Salmonella*. Frontiers in Microbiology. 2014; 5:52. doi: 10.3389/fmicb.2014.00052

[34] Eng SK, Pusparajah P, Ab Mutalib NS, Ser HL, Chan KG, Lee LH. *Salmonella*: a review on pathogenesis, epidemiology and antibiotic resistance. Frontier in Life Science. 2015; 8:284-293.

[35] Noda T, Murakami K, Etoh Y, Okamoto F, Yatsuyanagi J, Sera N. Increase in resistance to extended-spectrum cephalosporins in *Salmonella* isolated from retail chicken products in Japan. PLoS ONE. 2015; 10(2): e0116927. doi:10.1371/journal.pone.0116927.

[36] Liakopoulos A, Geurts Y, Dierikx CM, Brouwer MS, Kant A, Wit B, Extended-spectrum cephalosporin-resistant *Salmonella enterica* serovar *Heidelberg* strains, the Netherlands. Emerging Infectious Disease. 2016; 22: 1257. doi: 10.3201/eid2207.151377

[37] Wong MH, Chan EW, Liu LZ, Chen S. PMQR genes oqxAB and aac(6') Ib-cr Accelerate the development of fluoroquinolone resistance in *Salmonella Typhimurium*. Frontiers in Microbiology. 2014; 5:521. doi: 10.3389/fmicb.2014.00521

[38] Bearson BL, Brunelle BW. Fluoroquinolone induction of phage mediated gene transfer in multidrug-resistant *Salmonella*. International Journal of Antimicrobial Agents. 2015; 46: 201-204.

[39] Carattoli A, Tosini F, Giles WP, Rupp ME, Hinrichs SH, Angulo FJ. Characterization of plasmids carrying CMY-2 from expanded spectrum cephalosporin- resistant *Salmonella* strains isolated in the United States between 1996 and 1998. Antimicrobial Agents and Chemotherapy. 2002; 46 (5): 1269-1272.

[40] Guerra B, Soto S, Helmuth R, Mendoza MC. Characterization of a self- transferable plasmid from *Salmonella enterica* serotype *Typhimurium* clinical isolates carrying two integron-borne gene cassettes together with virulence and drug resistance genes. Antimicrobial Agents and Chemotherapy. 2002; 46: 2977-2981.

[41] Chen L, Chavda KD, Melano RG, Hong T, Rojtman AD, Jacobs MR. Molecular survey of the dissemination of two blaKPC-harboring IncFIA plasmids in New Jersey and New York hospitals. Antimicrobial Agents and Chemotherapy. 2014; 58: 2289 - 2294.

[42] Zhang Y, Lejeune JT. Transduction of blaCMY-2; tet(A), and tet(B) from *Salmonella enterica subspecies enterica* serovar *Heidelberg* to *S. Typhimurium*. Veterinary Microbiology. 2008; 129: 418-425.

[43] Rakhuba D, Kolomiets E, Dey ES, Novik G. Bacteriophage receptors, mechanisms of phage adsorption and penetration into host cell. Polish Journal of Microbiology. 2010; 59(3): 145-155.

[44] Smillie C, Garcillan-Barcia MP, Francia MV, Rocha EP, de la Cruz F. Mobility of plasmids. Microbiology and Molecular Biology Review. 2010; 74: 434-452.

[45] Banuelos-Vazquez LA, Torres-Tejerizo G, Brom, S. Regulation of conjugative transfer of plasmids and integrative conjugative elements. Plasmid. 2017; 91: 82-89.

[46] Lawley TD, Klimke WA, Gubbins MJ, Frost LS. F factor

conjugation is a true type IV secretion system. FEMS Microbiology Letter. 2003; 224: 1-15.

[47] Wong JJW, Lu J, Glover JNM. Relaxosome function and conjugation regulation in F-like plasmids— a structural biology perspective. Molecular Microbiology. 2012; 85: 602-617.

[48] Waksman G. From conjugation to T4S systems in gram-negative bacteria: a Mechanistic biology perspective. EMBO Rep. 2019; 20:e47012. doi: 10.15252/embr.201847012

[49] Le Hello S, Bekhit A, Granier S A, Barua H, Beutlich J, Zając M, Münch S, Sintchenko V, Bouchrif B, Fashae K, Sontag L, Fabre L, Garnier M, Guibert V, Howard P, Doublet B, Weill F-X. The global establishment of a highly fluoroquinolone resistant *Salmonella enterica* serotype *Kentucky* ST198 strain. Frontiers in microbiology. 2013; 4: 395 – 399.

[50] Beutlich J, Guerra B, Schroeter A, Arvand M, Szabo I, Helmuth R. Highly Ciprofloxacin resistant *Salmonella enterica* serovar *Kentucky* isolates in turkey meat and a human patient. Berlin and Munchen Tierarzotil Wochenscher. 2012;125: 89 – 95.

[51] Münch S, Braun P, Wernery U, Kinne J, Pees M, Flieger A. Prevalence, serovars, phage types, and antibiotic susceptibilities of *Salmonella* strains isolated from animals in the United Arab Emirates from 1996 to 2009. Tropical Animal Health Production. 2012; 44: 1725– 1738.

[52] Wasyl D, Hoszowski A. First isolation of ESBL-producing *Salmonella* and emergence of multiresistant *Salmonella Kentucky* in turkey in Poland. Food Research Institute. 2012; 45: 958-961.

[53] Barua H, Biswas PK, Olsen KEP, Shil SK, Christensen JP. Molecular Characterization of motile serovars of *Salmonella enterica* from breeder and commercial broiler poultry farms in Bangladesh. PLoS ONE. 2013; 8(3): 1 – 9.

[54] Fashae K., Leekitcharoenphon P, Hendriksen RS. Phenotypic and Genotypic Comparison of Salmonellae from Diarrhoeic and Healthy Humans and Cattle, Nigeria. Wiley Zoonoses and Public Health. 2017; 65:185-195.

[55] Sharma CS, Dhakal J, Nannapaneni R. Efficacy of lytic bacteriophage preparation in Reducing *Salmonella* In Vitro, on Turkey Breast Cutlets, and on Ground Turkey. Journal of Food Protection. 2015; 78(7): 1357-1362.

[56] Hawkey J, Le Hello S, Doublet B. Global phylogenomics of multidrug - resistant *Salmonella enterica* serotype *Kentucky* ST198. Microbial genomics. 2019; 5: 1-12.

[57] Afema JA, Byarugaba DK, Shah DH, Atukwase E, Nambi M, Sischo WM. Potential Sources and Transmission of *Salmonella* and Antimicrobial Resistance in Kampala, Uganda. PLoSONE. 2016; 11(3): 1 – 21.

[58] Raufu IA, Fashae K, Ameh JA, Ambali A, Ogunsola FT, Coker AO, Hendriksen RS. Persistence of fluoroquinolone-resistant *Salmonella enterica* serovar *Kentucky* from poultry and poultry sources in Nigeria. Journal of Infection in Developing Countries. 2014; 8(3): 384-388.

[59] Ziyate N, Karraouan B, Kadiri A, Darkaoui S, Soulaymani A, Bouchrif B. Prevalence and antimicrobial resistance of *Salmonella* isolates in Moroccan laying hens farms was undertaken. The Journal of Applied Poultry Research. 2016; 25 (4): 539 – 546.

[60] Allaoui AE, Rhazi Filali F, Ameur N, Bouchrif B. Contamination

of broiler turkey farm by *Salmonella* spp. in Morocco: prevalence, antimicrobial resistance and associated risk factors. Scientific and Technical Review. 2017; 36 (3): 1 – 30.

[61] Harrois D, Breurec S, Seck A, Delaune A, Le Hello S, Pardos de la Gandara M, Sontag L, Perrier-Gros-Claude JD, Sire JM, Garin B, Weill F-X. Prevalence and characterization of extended-spectrum b-lactamase-producing clinical *Salmonella enterica* isolates in Dakar, Senegal, from 1999 to 2009. Clinical Microbiology. Infection 2013; 20: 109 – 116.

[62] Welch TJ, Fricke WF, McDermott PF, White DG, Rosso ML, Rasko DA. Multiple antimicrobial resistance in plague: an emerging public health risk. PLoS One. 2007; 2:e309. doi: 10.1371/journal.pone.0000309

[63] Hoffmann M, Pettengill JB, Gonzalez-Escalona N, Miller J, Ayers SL, Zhao S. Comparative sequence analysis of multidrug-resistant IncA/C plasmids from *Salmonella enterica*. Frontiers in Microbiology. 2017; 8:1459. doi:10.3389/fmicb.2017.01459.

[64] Cao G, Allard M, Hoffmann M, Muruvanda T, Luo Y, Payne J. Sequence analysis of IncA/C and IncI1 plasmids isolated from multidrug-resistant *Salmonella Newport* using single-molecule real-time sequencing. Foodborne Pathogen and Disease. 2018; 15: 361-371.

[65] Johnson TJ, Thorsness JL, Anderson CP, Lynne AM, Foley SL, Han J. Horizontal gene transfer of a ColV plasmid has resulted in a dominant avian Clonal type of *Salmonella enterica* serovar *Kentucky*. PLoS One. 2010; 5(12): 1 – 10.

[66] Villa L, Garcia-Fernandez A, Fortini D, Carattoli A. Replicon sequence typing of IncF plasmids carrying virulence and resistance determinants. Journal of Antimicrobial Chemotherapy. 2010; 65 (12): 2518-2529.

[67] Chen K, Dong N, Zhao S, Liu L, Li R, Xie M. Identification and characterization of conjugative plasmids that encode ciprofloxacin resistance in *Salmonella*. Antimicrobial Agents and Chemotherapy. 2018; 62: 575– 618.

[68] Han J, Lynne AM, David DE, Tang H, Xu J, Nayak R. DNA sequence analysis of plasmids from multidrug resistant *Salmonella enterica* serotype *Heidelberg* isolates. PLoS One. 2012; 7:e51160. doi: 10.1371/journal.pone.0051160

[69] McMillan EA, Gupta SK, Williams LE, Jové T, Hiott LM, Woodley TA. Antimicrobial resistance genes, cassettes, and plasmids present in *Salmonella enterica* associated with United States food animals. Frontiers in Microbiology. 2019; 10:832. https://doi.org/10.3389/fmicb.2019.00832

[70] Chen W, Fang T, Zhou X, Zhang D, Shi X, Shi C. IncHI2 plasmids are predominant in antibiotic-resistant *Salmonella* isolates. Frontier in Microbiology. 2016; 7:1566. https://doi.org/10.3389/fmicb.2016.01566

[71] McMillan E A, Jackson C R, Frye J G. Transferable Plasmids of *Salmonella Enterica* Associated With Antibiotic Resistance Genes. Frontiers in Microbiology. 2020;11:562181. doi: 10.3389/fmicb.2020.562181